NEW HISTORIES OF
SCIENCE, TECHNOLOGY, AND MEDICINE

Series Editors

Margaret C. Jacob

Spencer R. Weart

Harold J. Cook

PRACTICAL MATTER

Newton's Science in the Service of Industry and Empire, 1687–1851

Margaret C. Jacob and Larry Stewart

Harvard University Press

Cambridge, Massachusetts

London, England

2004

Library of Congress Cataloging-in-Publication Data

Jacob, Margaret C., 1943–
Practical matter : Newton's science in the service of industry
and empire, 1687–1851 /
Margaret C. Jacob and Larry Stewart.
p. cm. — (New histories of science, technology, and medicine)
Includes bibliographical references and index.
ISBN 0-674-01497-9 (alk. paper)
1. Newton, Isaac, Sir, 1642–1727.
2. Science—Philosophy—History—17th century.
3. Science—Philosophy—History—18th century.
4. Science—Philosophy—History—19th century.
5. Science—History—17th century.
6. Science—History—18th century.
7. Science—History—19th century.
I. Stewart, Larry, 1946–   II. Title.   III. Series.

Q174.8 .J33 2004
501   2004052293

This book is dedicated to Trevor Levere,
for his inspiration and his kindness

Contents

PRACTICAL MATTER

Introduction

In the 1690s in all the Italian cities where the Inquisition operated, repression of thought deemed heretical rose to a crescendo. People stood trial for beliefs that had surfaced decades earlier. Since the 1630s and the time of Galileo, the Roman Catholic Inquisition had accused the Church's enemies of promoting atomism and other doctrines associated with the new science. It believed that the notion of atoms sanctioned the triumph of blind matter over spirit, hence of atheism over trust in the power of God. If all nature was formed through the result of the blind action of the minuscule particles of matter, then life, even God, would be replaced by nothing but matter and motion. As the prosecutor at the Venice trial said, if "the first man was composed of atoms like all other animals, everything resides in nature, God does not exist and neither does hell, purgatory or heaven, and the soul is mortal."[1] These were powerful arguments against the adoption of the new science, and believing in atomism could land someone in prison, or worse. In such a religious climate, how could matter, if defined by the new science as atomic, be embraced as having all sorts of practical and potentially progressive applications?

The innovative natural philosophy that so appealed to Galileo was heliocentric and atomistic. It had arisen in the seventeenth century partly inspired by Copernicus's claim, first made in 1543, that the sun was at the center of the universe. Many factors worked to make the Copernican model of the heavens attractive. Galileo's experiments with falling bodies and his discoveries made through the telescope seemed to suggest that all of nature is uniform. If Saturn has moons that orbit around it, might not the earth orbit the sun? Copernican mathematics was more elegant and Copernicus's system actually simpler than its geocentric rival, which was first proclaimed in the second century by Ptolemy. But more was at work here than just new discoveries. The ideas that so vexed the Church and led it to put Galileo under house arrest, and two generations later to instigate the trials in the 1690s, occurred at a time of great social and economic change, the result of the enormous commercial expansion experienced throughout the Western world.

New and distant places, recently discovered, licensed new ideas, the experimental exploration of nature, and even new approaches to mathematics. Natural philosophers argued that the great discoveries made by exploration licensed new philosophies. Indeed, a recent historian has argued that overseas exploration unleashed new and compelling metaphors for change, for seeking and actually finding the previously unknown: "The imagery of mathematics as a voyage of discovery was closely associated with the development of the new and controversial infinitesimal techniques."[2] In the new mathematical practices of the day, the space of geometrical forms came to be divided into increasingly small units, and their relation to one another could explain the symmetry of the whole of nature. With such a method, rather than with the old deductive system of Euclidian geometry, space came to be seen as capable of infinity, no longer simply something occupied by squares and triangles. Bodies moving through such space suggested that their trajectories should be plotted. Turning away from Euclid led ultimately to the calculus—simultaneously and independently discovered in the 1670s by Newton and Leibniz. The indivisible atoms could be imagined as moving in a continuum with knowable trajectories. In the seventeenth century, in the worlds celestial and terrestrial, everything seemed up for grabs; none of the old certainties about the land masses of our planet, or about the way space and bodies should be described, could be taken as given. All these changes were immensely threatening to anyone who valued order, orthodoxy, and assent to traditional authority. Much was at stake in the Italian trials of the 1690s.

This book examines why and how, despite the opposition of the Inquisition and countless other powerful guardians of tradition, a scientific understanding of the world gained acceptance and application throughout much of the Western world. The book also places particular emphasis on Great Britain, where, before many other places, science was made practical and put into the service of industry and empire. The science that proved so fertile relied on a set of interlocking assumptions put in place gradually in the course of the seventeenth century. Matter was composed of indivisible atoms, some said corpuscles; all motion in the universe resulted from one body's impact upon another; rest meant simply the absence of contact between bodies (not some preexisting state of inertia); and all the bodies seeable by the human eye, from planets to the objects newly revealed by the microscope, partook of the same physical matter, differently configured. The configurations and motions could be expressed mathematically, particularly by the introduction of the new algebraic symbols joined to the method of indivisibles. Last, but far from least, the sun was at the center of the universe. Now conceived as physical bodies, the planets revolved about the sun, not the earth. Any one of these assumptions might be said to contradict the Bible, and especially the writings of the ancient philosopher Aristotle, as reworked by his medieval interpreter St. Thomas Aquinas. The potential for controversy and, in the eyes of the Church, heresy, lurked everywhere.

For reasons discussed in Chapter 1, in the 1690s Protestant, but not Catholic, Europe became more congenial to the new science.

Just when the Italian heretics were being tried, hundreds of miles to the north, English Protestant clergymen rose in their London pulpits to explain atoms and invisible forces to their well-heeled congregations—all to assist their coming to terms with Newton's scientific laws. In general, south of the Alps in Catholic Europe, science found fewer comfortable settings than in the cities of England, the Low Countries, France, and a few places in Germany. Even in St. Petersburg, Russia, new ways of thinking about nature could be found in books, and most importantly, at learned societies and academies where experiment and discussion flourished. So too in the Atlantic colonies of Europe, from Boston to Saint Dominique (today Haiti), new ways of thinking about nature traveled with educated European settlers and traders. Where the Inquisition had little influence, gradually atomism became commonplace and so too did the assumption that the sun lay at the center of the infinite universe. This was true even in Catholic France where, outside of the Papal territory of Avignon, the king would not allow the Inquisition to exercise its power.

Early in the seventeenth century, Francis Bacon, the English statesman with an eye toward the practical, pleaded with the educated to imitate the voyagers, to explore nature, to collect, classify, and experience. As his writings were translated into every European language, the Baconian message spread. By 1798, when the French revolutionaries opened the first-ever industrial exposition in Paris, the minister in charge invoked the memory of Francis Bacon. Lecturing just a few years earlier in the German university town of Jena, the professor of physics introduced the course with a brief history of science that began with the ancient Greek philosophers, Plato and Aristotle, and then jumped to all the major books by Bacon and Isaac Newton.[3] In the 1790s the once-embattled new science had achieved ascendancy from Philadelphia and Dublin to Berlin and Moscow. In the mid-nineteenth century Hebrew writers borrowed from these many explications of the new science and gave a quick history of science. As you will see at the opening of Chapter 2, they also jumped straight from the ancients to Galileo, as if the Middle Ages and the Renaissance had never existed. The Hebrew writers used scientific learning to help make their language more secular and less exclusively for use in explicating the sacred. By 1800 the race was on to test the usefulness of any science. Entrepreneurs and industrializing nation states benefited from the ensuing profits as mechanical principles and experimental habits were applied to mining, manufacturing, and transportation. The race for industrial development fueled the contest for empire.

By 1800 British industrial innovation and expansion set the economic pace in the West. Fired by their own nationalism, Continental Europeans worried obsessively about their ability to compete with developments across the Channel, to be found in the factories of Birmingham and Manchester. Early in the nineteenth century aspiring British industrial entrepreneurs, needing to understand steam engines, owned manuals that taught mechanics as well as the political economy of Adam Smith. In piously Protestant Britain, cotton mill owners could believe that "the greater the improvement in machinery or any other science . . . then the greater good to all human beings."[4] But in the factories, their badly paid workers might not have been so convinced that their condition represented progress.

In the Western world during the nineteenth century, a faith in capital and industry emerged and appeared unshakeable: "We are now more industrious than our forefathers, because our capital, destined for the encouragement of industry, is greater."[5] Science wedded to technology—their union sanctioned by capital investment—had produced new wealth and techniques that would revolutionize human productivity and eventually, slowly, even the wealth and life expectancy of workers. The message of progress appealed to mechanically inclined artisans as well as to entrepreneurs with money. Along with lessons in geometry and mechanics, they read potted biographies of Newton, Bacon, and Benjamin Franklin, and they even learned of young artisans who transformed themselves from being conjurers into serious mathematics instructors.[6] Household manuals promoting applied sciences taught men and women everything from chemistry and the latest mechanical inventions to domestic economy, right down to how to get rid of bedbugs and cook potatoes.[7] Textbooks led to the habit of "looking up things" and the assumption that expert knowledge could improve life and promote prosperity. By the 1820s French engineers, in their haste to narrow the British lead, were being exhorted in their own manuals on mechanics to assist the needs of both the state and commerce. They were further instructed on the steam engine, "the greatest party to the riches of industrial England," an object, according to the manuals, not sufficiently understood in France.[8]

This book examines a profound transformation. Gradually from 1687, when Newton published his great book on mechanics and celestial dynamics, to 1851, at the largest industrial exhibition ever seen, which was held at the specially built Crystal Palace in London, science became central to Western thought and economic development. We look at the myriad ways and means science began to be understood, and in the process became so fundamental to Western culture, integrated and applied to everything from the study of the heavens, rocks, and plants to the making of industrial devices. The book is short and the topic vast. There is no space or inclination to look at technology *per se*, although one of our claims is that mechanical science as articulated by the British Newtonians had a profound impact on early industrial development. Our focus is almost entirely on the uses of mechanics, by far the most commonly taught and widely read form of post-*Principia* science. We examine its uses, its effect on the imagination, and ultimately on the wealth of Westerners. The rise to prominence of a science aimed at application accelerated markedly during the lifetime of Isaac Newton (1642–1727). Within a hundred years, the benefits derived from its application became incontrovertible as people marveled at the great industrial expositions of the nineteenth century. By the middle of that century, taking the famous exposition of 1851 as our closing point, of all the approaches to the varieties of nature, only medicine remained relatively "unscientific," but that too was changing very rapidly with the new chemistry and the discovery of germs and anesthesia.

In beginning our story with Newton and his legacy, we leave out the thousands of philosophical practitioners who preceded him; and in the opening chapter we reference his great predecessor Descartes only in passing. But we must start somewhere. Newton opened up an entirely new phase in the assimilation of science: Mechanical science based on his discoveries became the foundation for religious thought, for what Protestants called natural religion. Within a century of his death in 1727, the laws of Newtonian physics also provided the model for the mathematization of randomness. Laws of error and probability could be calculated by the new science of statistics, and pioneers like the Belgian Apolphe Quetelet said that even social change could be quantified. He articulated the mathematically supported concept of *l'homme moyen*—the average man who was "in a nation what the center of gravity is in a planet."[9] He claimed that the social—like the natural—could be known scientifically and statistically. Only in Chapter 2 do we, as it were, go backward and give some hint of the science that Newton had to overthrow, first that of Aristotle and even that of Descartes.

As Chapters 3 and 4 make clear, throughout the eighteenth and early nineteenth centuries, a vast army of lecturers, experimenters, engineers, schoolteachers, and professors made the new science accessible to practical goals and applications. Chapter 5 examines one place, Manchester, where applied science created a new industrial and social order. It also demonstrates the enormous efforts made by the French to catch up with British industrial development. By 1851, at the time of the great London exhibition of industry and science, a brave new world had come to pass, first in Britain, and it could be put on triumphant display. That Victorian setting was less fettered and more optimistic than the one later shattered by the world wars of the twentieth century, or after Hiroshima by the lethal by-products of atomic science. Science in the service of Western industry and imperial expansion seemed in 1851 so obviously to benefit those who possessed it. The have-nots lacked the power that came with such knowledge. Yet for all the injustice associated with Western imperial expansion, even with the efforts to make racial inferiority appear to have a scientific foundation—a favorite sport of American and European scientists of the mid-nineteenth century— there was nothing preordained or inevitable about the centrality awarded to science in the West. The contingency of the award, the steps along the way, require our attention. It is easy to forget that modern science might have been stillborn, or have remained the esoteric knowledge of court elites, or in a darker scenario, limited to the thinking of heretics persecuted whenever the authorities had the opportunity. Instead—for better and occasionally for worse—in the lifetime of Newton, matter turned practical and scientific knowledge became a centerpiece of Western culture, a partner with industry, war-making, and, in budgetary terms, the largest component of any modern Western university.

CHAPTER 1

The Newtonian Revolution

Born in 1642 in Lincolnshire in England, in the year Galileo died, Isaac Newton grew up in the midst of civil wars. He was a country boy whose father died before he was born and whose mother appears to have had little time for him once she remarried. At Cambridge University, he waited on student tables to pay his way. This boy with an unsettled passage became a philosopher and mathematician who revolutionized our understanding of the heavens. Yet probably his first love throughout his long, celebrated life—he died, world famous, in 1727—was theology. This man of the Bible, who believed that God would end the world and usher in a millennial paradise, all to be preceded by the conversion of the Jews, ironically did more than any other mortal to make the world seem like an ordered, rational, certain place, bounded by mathematically knowable laws that would stay in place forever.

The private Newton whose fantasies about the end of the world endlessly fascinate us today stayed largely hidden in his own lifetime, known only to a select few of his very small circle of friends.[1] We know that he wrote his most famous book, *Mathematical Principles of Natural Philosophy* (1687, hereafter, his *Principia*), to make humankind believe more deeply in the deity. But that purpose does not leap out when the reader first approaches the text. How readers prior to the mid-nineteenth century read the *Principia*, what they could take from it, synthesize, or rework, how its legacy entwined with other ideas and institutions—that process vitally concerns us in this and subsequent chapters.

At the foundation of Newton's science—and all subsequent Western science—rested one bedrock assumption. Put in Newton's own words, "nature is exceedingly simple and conformable to herself" and that means whatever "holds for greater motions, should hold for lesser ones as well."[2] The rules, Newton said in the *Principia,* were universal. In other words, the laws that hold for the physics of local motion, or mechanics, also hold for planetary motion, even for the movement of the invisible forces that Newton saw as electric and aethereal. With similar assumptions, neuroscientists in our day can work on the hippocampus in the brain of mice knowing that its molecular structure is almost indistinguishable from that found in the human brain. By analogy, they work on the first in order to elucidate the second.

Similarly, when Newton tackled the problem of why bodies on the surface of our planet are not thrown into space by its rotation and annual orbital revolution, he rigged up a swinging device that was analogous, "a conical pendulum 81 inches long, at an incline of 45 degrees to the vertical."[3] By measuring its swing and the hold that earthly gravity exacted on it, Newton could show how the centrifugal pressures on bodies on the planet held them in check despite the earth's rotation. The forces at work on a tabletop pendulum could also be extrapolated to the gravitational pull between planets. This example illustrates why Newton's style of reasoning and his examples organized mechanics of everyday bodies just as much as they did celestial mechanics. The principle of universal gravitation that holds the heavenly planets in their orbits around the sun also works on bodies here on earth.

In the *Principia* Newton provided a basic set of definitions: The mass, or quantity of matter, of a body is proportional to its weight. This axiom was proven by a set of experiments using identical pendula hung from cords of equal lengths to which were attached bodies of various sizes and substances but of equal weights. As they swung, the different bodies wound up at the same place at the same time. What this means in practice is that bodies compressed, or bent, or heated, or taken up a mountain, or out into space retain the same mass (although not the same weight as they move further from the earth). In the generation just prior to Newton, the great French natural philosopher René Descartes had defined mass as simply that which occupies space. For Newton and all his followers, this definition was too simplistic. For them the quantity of matter is measured by its density and magnitude, and the focus must be experimental or experiential if the researcher or engineer is to know how much mass is present at a site. In effect, density is used to mean the specific gravity of a body. Newton is demanding experimental proof, but he is also arguing and proving that mathematics can be applied to mechanics. As he put it in the preface to the *Principia*, "*geometry* is founded on mechanical practice."

Newton often said that he stood on the shoulders of giants. As we will see in greater detail in the next chapter, two generations of natural philosophers preceded him. Self-consciously Newton built on the mechanical experiments of Galileo and explained that, as early as 1610, Galileo "found that the descent of heavy bodies is the squared ratio of the time and the motion of projectiles occurs in a parabola, as experiment confirms, except insofar as these motions are somewhat retarded by the resistance of the air."[4] Again Newton illustrated the phenomenon by oscillating pendula and noted the analogy that it "is supported by daily experience with clocks." Newton aimed at every turn to provide universal, testable rules, "in the motions of machines those agents (or acting bodies) whose velocities . . . are inversely as their inherent forces are equipollent and sustain one another by their contrary endeavors." The principle by which force is increased by decreasing velocity can then be illustrated by resort to clocks and similar devices in which "the contrary forces that promote and hinder the motion of the gears will sustain each other if they are inversely as the velocities of the parts of the gears upon which they are impressed." Similarly such forces are at work when a hand turns a screw into wood and "the same is the case for all machines."[5] Newton could not have been more concrete: "If machines are constructed in such a way that the velocities of the agent or acting body and the resistant are inversely as the forces, the agent will sustain the resistance and, if there is a greater disparity of velocities, will overcome that resistance."

Just when the new principles might have become interesting for men who had been hauling and pushing all their lives—and the possibility of rationalizing what they did presented itself—Newton announced that he was after a far bigger, cosmic picture: "My purpose here is not to write a treatise on mechanics. By these examples I wished only to show the wide range and the certainty of the third law of motion," namely, all things being equal (which they seldom are), "the action and reaction will always be equal to each other in all examples of using devices or machines."[6] In two generations from Galileo to Newton, natural philosophy had moved from the study of local motion to the search for universal laws of motion at work in the heavens and on earth.

Incontrovertibly, Newton's purpose in writing the *Principia* was to provide the mathematical laws at work in the heavens, and he ended complex sections of his great work almost offhandedly: "So much for the finding of orbits. It remains to determine the motions of bodies in the orbits that have been found."[7] With a mathematical complexity that dumbfounded even the learned philosopher John Locke (who had to write to a friend in The Netherlands for assistance), Newton explicated the law of universal gravitation. Throughout the text, Newton proves mathematically that the force keeping the planets in elliptical, orbital motion around the sun operates from their centers uniformly as inversely the square of the radius being traversed. He postulates that the centripetal forces at work *are* attractions without ever telling his readers what attraction is. Famously, he dismisses that question by saying, "I do not frame hypotheses."

We now believe that he privately thought of universal attraction as the will of God operating in the universe. The force of gravitation acts on all the heavenly bodies through "action at a distance." "Wherever matter is, there gravity is also."[8] No mechanism, contact, or push-pull encounter between bodies, however minuscule, is needed to effect the attraction and repulsion that holds the universe in place. For purposes of his mathematics, Newton postulated a vacuum in space through which the attractive forces act. But gravity is not *inherent or essential* to matter. For reasons that were deeply religious, Newton could not imagine a universe possessed of the power to move itself without recourse to its supernatural source. Left to its own devices, matter possesses only inertia or, as Newton put it, matter is "brute and stupid." Newton, the atomist, rejected the notion that the doctrine eliminated supernatural life or will from the universe. Perhaps the judges in the Naples trials of the 1690s should have brought him over and taken his testimony. But just as the Italian judges feared, over the course of the eighteenth century many of Newton's readers found no need for such a pious prohibition against force as inherent in the nature of matter.

Historians often repeat the tale about Newton being spotted by a couple of students in the lanes of Cambridge before he moved to London. One undergraduate says to the other, "There goes the man who wrote that book that no one can understand." This has a ring of truth to it, if only because the newly invented calculus as well as the geometry of the *Principia* was difficult even for those well versed in mathematics. But Newton also promoted elusive philosophical principles like attraction across an utterly void space— thus worked gravity or even magnetism—that mechanical theory had long thought imaginary. If the witty undergraduate accurately reflected the considered Cambridge opinion of Newton, then how could it be that his philosophy was not simply dismissed or ignored? Why did it create not merely a great debate but also a widespread admiration by the end of his life? In 1727, following the honor of a state funeral, among great statesmen and poets, Newton was buried in Westminster Abbey.

Newton also wrote a more readily intelligible and nonmathematical work on the nature of light. The publication of his *Opticks,* first in 1704 and then in two subsequent and enlarged editions, put in circulation a much more approachable book for the educated reader than the forbidding *Principia.* But one relatively easy book can hardly explain Newton's reputation. Nor is it sufficient to claim that Newton's remarkable mathematical brilliance suffices as a justification for the enthusiasm with which the British, at the very least, embraced his philosophy. One of the most striking facts of the early eighteenth century was the way Newtonian science achieved public status even when Newton himself deliberately distanced himself from those whom he called "vulgar smatterers in mathematicks."

An examination of Newton's reputation throughout the eighteenth century ultimately reveals a strain between, on the one hand, a wide and growing public acceptance of science and, on the other, a converse reliance on exclusivity in a culture long defined by rank and social status. At least since the time of the Elizabethan courtier Francis Bacon, there had been a vexed issue about the role of craftsmen and artisans in the natural philosophical world populated by gentlemen and university scholars. That tension is displayed in the distinction drawn in Elizabeth's reign by the compass maker Robert Norman between what he then saw as the imaginative "conceits" of learned scholars and the worldly experience of mechanics and mariners. In other words, running through the history of the scientific revolution from Bacon to Newton lies a nearly instinctive divergence of interests between practical utility and gentlemanly curiosity. The practical and the artisanal were segregated from the genteel by social standing.[9] That segregation is explored more thoroughly in Chapter 3. Right now we need to address how, despite this segregation, Newton's science achieved relatively early success in England and Scotland. In those places after 1700, the paradigm decisively shifted away from Aristotle and Descartes toward Newton as the authoritative guide to physical nature.

Because Newton's achievements in celestial mechanics were so spectacular, historians have invariably focused on his discussions of this topic, particularly in Books 1 and 3 of the *Principia*. But in this book we wish to read the text as did many of Newton's immediate followers, and most of his explicators for at least three generations into the early nineteenth century. They focused on the mechanics of earthly bodies and, in so doing, aided immensely in the triumph of Newtonian science. They found in Book 2 of his masterpiece the foundations for the study of fluids in motion and at rest, of hydrodynamics and hydrostatics. The *Principia* made extensive use of experimental evidence. Its appeal to engineers became immediately obvious—indeed, the next generation of Newtonian engineers, including men like John Smeaton, invented civil engineering as a profession and distinguished it from military engineering. In alliance with entrepreneurs, often themselves schooled in applied Newtonian mechanics, the engineers laid the foundations for early industrial development in Britain. The *Principia* will always remain a great book, possibly the greatest ever published in science. But it was the practitioners, the audience, the new public, the buyers and consumers of the new science, who made it the cornerstone of Western economic development.

Book 2 of the *Principia* reveals its practical side, while Book 3 features propositions that were intended to popularize and simplify Newton's discoveries. In so doing, Newton himself may be said to have laid down a template for what became in the eighteenth century a vast industry of Newtonian textbooks. Like Book 3, the textbooks moved from generalization about the heavens to experiments on earth that illustrate the principle. For example, Newton tells his readers "the falling of all heavy bodies toward the earth (at least on making an adjustment for the inequality of the retardation that arises from the very slight resistance of the air) takes place in equal times, and it is possible to discern that equality of the times, to a very high degree of accuracy, by using pendulums. I have tested this with gold, silver, lead, glass, sand, common salt, wood, water and wheat."[10] Newton rigged up two boxes of equal weight and, in the center of each, put different substances that weighed the same. The boxes were suspended from eleven-foot cords and put in oscillation. They moved in exact symmetry, and Newton concluded that "there is no doubt that the nature of gravity toward the planets is the same as toward the earth."[11]

Next comes a discussion of the satellites of Jupiter and how "in equal times in falling from equal heights toward Jupiter they would describe equal spaces, just as happens with heavy bodies on this earth of ours." Corollaries followed: The weights of bodies do not depend on their forms and textures and "the weights of all bodies that are equally distant from the center of the earth are as the quantities of matter in them. This is a quality of all bodies on which experiments can be performed."[12] Always Newton turns the reader's attention toward the heavenly planets and explains, for instance, that planets closer to the sun are denser in their matter. If our planet were in the orbit of Mercury, water "would immediately go off in a vapor." As we now in 2004 find traces of water on Mars, we are still operating with assumptions about what the planets must be like that first appeared in the *Principia*.

Reading the range of measurements available to Newton, from the seconds of a pendulum swing in Paris versus its length of arc on the islands of Gorée, Guadeloupe, and Martinique, and elsewhere in the Americas and Africa, we realize that Newton's achievement can be tied to the vast increase in general knowledge that overseas trade and exploration had brought to Europeans.[13] The courage to generalize, to arrive at universals about the natural world, owes much to the immense quantity of information—and self-confidence—that European mastery of the seas gave land-bound thinkers like Isaac Newton. Measurement of tidal changes in the seas of the world made it easier to be precise about the moon's effect on the tides, a subject also elucidated in Book 3 of the *Principia*. It ends with a discussion of comets and a demonstration that within the framework of their apparent eccentricity lay traceable, hence knowable, orbits. Only when the labor of natural philosophy and mathematics had been laid to rest did Newton turn to the meaning of it all.

At the end of the *Principia*, Newton placed a General Scholium that was deeply religious in purpose and tone: "The most elegant system of the sun, planets, and comets could not have arisen without the design and dominion of an intelligent and powerful being . . . He rules all things, not as the world soul but as the lord of all. And because of his dominion he is called Lord God *Pantokrator* [that is, universal ruler]."[14] One more time, Newton refused to speculate about the exact nature of gravity. Instead he set forth the principles on which all science should proceed, and first among them: "In this experimental philosophy, propositions are deduced from the phenomena and are made general by induction. The impenetrability, mobility, and impetus of bodies, and the laws of motion and the law of gravity have been found by this method. And it is enough that gravity really exists and acts according to the laws that we have set forth."[15]

The *Principia* was an exercise of deep devotion to the Deity, a paean of praise that made nature the servant of God's will and power. Once explicated, nature's mathematical order, Newton believed, would bring humans to their knees in adoration of the creator. In part, that was wishful thinking. The entire trend of the century after Newton's death in 1727 was toward the secular, the here and the now, the enjoyment of this world, and the lessening of interest in the next. The trend was far from clear, however, when in the 1690s English clergymen rose in their pulpits and, in an unprecedented step, took up Newton's natural philosophy for the purpose of shoring up Christian belief.

Newtonian Science, Politics, and English Protestantism

Isaac Newton countenanced few close friends. One of them, the Anglican cleric and apologist Samuel Clarke, became a leading British intellectual of the early eighteenth century. During the highly unstable period of English politics ushered in by the Revolution of 1688–1689, Clarke lectured in the great London pulpits of St. Paul's and St. Martin-in-the-Fields on matters directly germane to the political situation. He lived in a setting that had been destabilized by the "rage of party," by factionalism between Whigs and Tories, and by perceived threats to the legal and moral status of the Anglican Church. In addition, treason lurked in the ranks of extreme Tories, embittered supporters of the exiled and defeated king, James II. Clarke was a supporter of the Revolution Settlement and a churchman willing to countenance a limited degree of non-Anglican Dissent from the official doctrines of the Church. Significant to the political setting, Clarke's lectures legitimated human liberty and the necessity for change while asserting the need for harmony and uniformity. Armed with the lessons he derived from his understanding of Newton's science, Clarke sought a *via media,* one that gave its blessing to the revolution but not to the republican, freethinking, and reformist elements that it unleashed. James II would have imposed both absolutism and Catholicism, and Anglicans like Clarke supported the revolution of 1688–1689 that deposed him and brought a new Protestant king, William III, to the throne by vote of Parliament. The events of that time period also saved the Church of England as established by law; and with it an Anglican establishment, both lay and clerical, retained political dominance.

Clarke's most famous set of lectures, delivered in an Anglican church in 1704, was entitled *A Demonstration of the Being and Attributes of God: More Particularly in Answer to Mr. Hobbs, Spinoza, and Their Followers.*[16] The lectures began by asserting the usefulness of theism. These were Clarke's Boyle lectures, an annual series endowed by the last will and testament of the great natural philosopher Robert Boyle. From their outset in 1692, the Boyle lectures were monopolized by Newtonians like Richard Bentley, William Derham, William Whiston, and, most famously, by Clarke. The lectures seem an unlikely vehicle for a meditation on the foundations of political order and stability. But in the clerical mind, stability required Protestant orthodoxy and the defeat of freethinking. More than any other faction within the Anglican Church, the Newtonians labored to bring Newton's science into service against atheism, more precisely against the materialism associated with Spinoza, Hobbes, and the alive and well John Toland. In response to the revolution of 1688–1689 and the salvation it offered to Protestantism in England, Anglican clerical ambition entailed laying a new foundation for political order and its handmaiden, religious orthodoxy.

Order and orthodoxy came to rest on an unprecedented meditation on the law-like, but divinely instituted, structure of the physical universe. The key to the success of this new theology that rested upon Newton's physics lay in theism. People had to believe that only a divine will could have established the laws of celestial mechanics as Newton had discovered them. In his lectures Clarke banished doctrines that asserted the independence of the material by noting that "if there be a vacuum as Newton maintained, it follows plainly that matter is not a necessary being . . . If an atheist will yet assert that matter may be necessary, though not necessary to be everywhere, I answer this is an express contradiction. For *absolute* necessity is absolute necessity *everywhere alike.*"[17] Hence, many hours into his demonstration of God's existence, Clarke asserts, "it is impossible there should be two different self-existent independent principles as some philosophers have imagined, such as God and matter."[18] Clarke equates independence of movement with human liberty; and once communicated by God to "created beings," they too have the freedom to exercise their will "in its proper place."[19] To deny free will, Clarke maintains, is to deny the human power to effect meaningful change, to begin a motion that "is a plain instance of liberty."[20] Without the will to move themselves, human beings become brute matter.

The following year in 1705, speaking from the pulpit of St. Paul's Cathedral, Clarke drew out the full implications of human liberty and the necessity for lawfulness: "The frame and order of the world is preserved by things being disposed and managed in a uniform manner."[21] And in case anyone missed the political implications of preserving things in a uniform manner, Clarke said it plainly: "Even the greatest enemies of all religion, who suppose it to be nothing more than a worldly or state-policy, do yet by that very supposition confess thus much concerning it. . . . For the practice of moral virtue does as plainly and undeniably tend to the natural good of the world; as any physical effect or mathematical truth, is naturally consequent to the principles on which it depends, and from which it is regularly derived."[22] Clarke's theism was as old as Christianity, but his recourse to the "laws of gravitation," the motion of the planets on their axes uniformly east to west, seemed to afford incontrovertible proof derived from Newton of the divine will at work in the universe. Clarke and his fellow Newtonians effected the marriage of two discourses, one scientific and the other religious, and their union was meant to be *politique,* to render the new constitutional structure into a providentially guided set of events. More than any other single factor in the rise to prominence of Newton's science, the postrevolutionary politics of the 1690s set the stage for its acceptance. Newton's science did ideological work in shoring up belief in a broad, liberal Christianity and in the providential order of a state sanctioned not by the divine right of kings but by a vote in Parliament. Not surprisingly, the leading Newtonian of the 1720s and 1730s, Jean Desaguliers, could speak about the "Newtonian System of Government."

Before Clarke rose in his pulpit, earlier English clergymen had sought to effect a marriage between science and religion. During the restoration of the monarchy following the civil wars of the 1640s and 1650s, Thomas Sprat, writing on behalf of the newly founded Royal Society, described experiment as a means of challenging the idolatry associated with radical sects. The hard work of experiment he and his Royal Society friends thought would suppress the tendency to sectarian madness that many saw at the root of the civil wars. Religious zeal had to be tamed and orthodoxy sustained. They initiated several decades of preaching and writing aimed at making science pious and religion, more scientific.

A generation later, Clarke's exact contemporary and close friend of Newton, William Whiston, made his reputation on speculations about the early state of the earth and the intercession of comets in his *New Theory* of 1696. One critic described his "snuff of a comet" as a hardly credible explanation of alterations in the earth's orbit and the creation of seasons.[23] But early in his career, Whiston was simply attempting to assert that physical evidence sustained the Christian account of creation. Hence, when he took employment in the Whig Richard Steele's "Censorium" in London, part of his mission as a public lecturer was to challenge those sectarians who believed that they had personal access to the true meaning of scriptural prophecy, a sentiment labelled "enthusiasm" and one that befuddled British politics before and during the civil wars.

Both Whiston and Clarke did battle against enthusiasm. It surfaced again in 1715 when the Protestant succession was contested in Britain as the crown passed, on the death of Queen Anne, from Stuarts to Hanoverians. Disturbances in the streets followed strange portents in the skies over early eighteenth-century London. Contemporaries related the prophetic signs to the uncertain survival of the new German monarchy. And new Protestant sects like the French Prophets took full advantage to read into eclipses, meteors, and northern lights over London alarming fears of Catholic assaults on the Protestant throne. Thus, the Newtonian lectures of Whiston and the pamphlets of the astronomer Edmund Halley ridiculed the sectarian pretenses that provoked unrest in the streets. A generation earlier, Sprat, it turned out, had been right about the usefulness of philosophers. The precarious nature of the Hanoverian succession in 1715 made the public lectures on astronomical phenomena seem more urgent. The Newtonian lecturers bent the weight of their science against "enthusiasm" and against prophets generally from the lower classes who thought they could "read" nature's signs by intuition alone.

Yet not everyone moved by the pious sense of order inspired by Newtonian science wanted just to sit in church pews. This was the age of pubbing and clubbing, and there was also a felt need to get to know men different from one's self. Into this propitious setting came the new Masonic lodges, which evolved out of the guilds of stonemasons. The Newtonian involvement in the new lodges was distinctive. Jean Desaguliers had unbounded energy and something of a common touch. By 1720, if not earlier, his imagination was captured as freemasonry sprang up in Scotland and England. Once entirely the precinct of working stonemasons, the lodges had begun to admit local men with an interest in architecture or investment in building projects. The lore of the workers had medieval roots and emphasized the unravelling of the mysteries of nature. The new science of Desaguliers fitted the aura of ancient wisdom rediscovered and a powerful, new form of socializing began. By 1730 it had been

1. Jean Desaguliers. One of the first to put Newtonian mechanics to work, he helped lay the foundations for a new knowledge-based economy in eighteenth-century England.
Courtesy of the Huntington Library, San Marino, California

transmitted to the Continent and by 1735 to Philadelphia, brought there from London by Benjamin Franklin.

Throughout the eighteenth century, thanks to the efforts of Newtonians like Desaguliers, Franklin, and the antiquarian William Stukeley, the Masonic lodges became associated with progress in science and the advocacy of the modern. Outside of Scotland, stonemasons disappeared from the lodges as aristocrats became Grand Masters and professional men (on the Continent, also women) became brothers. Some lodges even sponsored lectures in mechanics and natural philosophy. By midcentury, there were probably about 50,000 Freemasons in Europe, of which several hundred were women. They worshiped the Grand Architect, the God of the new science, and linked an interest in science with cosmopolitan fraternizing. The lodges became one of the few sites where the social gap between the elite and the middling and professional classes could be bridged. In some Scottish lodges literate artisans also belonged (indeed, they had started the Masonic movement); and by the middle of the eighteenth century, they were being given admission tests in mathematics and architecture by their more learned brothers.[24] The lodges allowed men of many different faiths to express a more general belief in the God of Newtonian order.

Throughout the eighteenth century, the British ideology that tied Protestantism to science, and used both to argue for order in society and government, had a European appeal. Decade by decade, Clarke's stature rose. He became more than just another smart English curate, a heresy hunter read by other curates on Saturdays when they were desperate to find something interesting to say the next morning. His Boyle lectures of 1704 went through seven editions in his lifetime, and they were translated into several European languages. As late as the 1760s, the great French political philosopher Jean-Jacques Rousseau ignored Clarke's polemics, as well as his deeply hierarchical politics, in order to use his natural religion for other purposes. Rousseau cited Clarke's lectures as laying the foundation for the natural religion that he would have his Savoyard vicar preach to maintain the peaceful and democratic state. In the next decade the archenemy of Christianity, the French atheist Baron d'Holbach, charged after Clarke as if he were alive and well, plagiarizing into French in 1770 a portion of a work by Clarke's enemy from the left, an English republican active in Newton's lifetime, John Toland.[25] Natural theology in the course of the eighteenth century had many uses, but all were intended to promote order and harmony. Society should mirror the stability of the heavens as revealed in Newton's *Principia,* and the promoters of change and reform from Toland to d'Holbach wanted to unsettle the established churches and the clergy that championed them.

Allied to Christian orthodoxy, by the 1720s Newtonian science in the hands of its many explicators was poised to effect a massive paradigm shift in the way Westerners thought about and related to nature. Experiment promised incremental knowledge, and mathematics had proven the possibility of laws. What no one foresaw—except the handful of first-generation Newtonians who pioneered applications in engineering—was how the new ways of thinking about nature would have economic consequences.

In the next chapter we move back in time, before Newton, and see how his precursors prepared the way for the triumph of Newtonian science. We examine how the paradigm shifted decisively in the direction of applied mechanical science of Newtonian inspiration. We also need to explore the nature of the public who emerged around science, who listened to its teachers, who bought the textbooks, and, most importantly, who met with groups of like-minded and commercially talented men who spied in the new knowledge a way to make a living, even a way to become rich and famous.

CHAPTER 2

The Western Paradigm Decisively Shifts

By the middle of the nineteenth century, Hebrew-speaking Jews in central and eastern Europe had developed strong interests in science, and they sought to make scientific learning accessible in the language of their communities. We may treat these Hebrew texts as if they were written by anthropologists who want to not only describe a different culture accurately but also translate it into terms familiar and comfortable. Russian Jews also formed a society "to broaden the education of Israel in Russia."[1] This effort at scientific education entailed explaining experimental physics, cosmology, astronomy, meteorology, chemistry, and some basic mathematics. All the information was derived from other textbooks and framed to illustrate "the wisdom of the Creator." The scientists mentioned were French and English, and the Hebrew texts even ended with a brief survey of time in the history of science up to the electrical experimenter Michael Faraday. These texts serve as testimony to a profound paradigmatic shift in the way Western cultures understood and approached physical nature.

In the Hebrew texts concepts that are absolutely basic to all science—atomism, force, momentum, gravity, equilibrium, the porosity of bodies, and the weight of air—are intermingled with drawings of machines. A Hebrew reader of one of these texts would have received a good account of the basic physics of the day and, by mastering that information, could have developed a reasonable scientific literacy. One Hebrew text emanating from Berlin in 1858 extolled the life of Alexander von Humboldt and explained that now "mankind has broken the boundaries of time and place and expanded its knowledge in every direction."[2] The terms used for von Humboldt are reverential; and when his work in geology is discussed, the language becomes almost Biblical: "The science of geology is like the book of Chronicles . . . in the layers of the earth are written with the hand of God the history of the world . . . and the scientist is reading this book of God and understands its signs."[3] By the 1850s in every Western language and in every urban corner—from St. Petersburg to Chicago—mechanical science and its ever-growing branches dominated the received wisdom of the learned.

This chapter outlines the beginnings of this process. It pauses in the 1730s, when in Britain Newtonian physics was tied to a Protestant (as well as Masonic) form of Christianity, and ends with the French Revolution. The shifting of the Western paradigm to a mechanical model of nature commenced in the seventeenth century, though in Chapter 1 we began our story not with Descartes, who died in 1650, but with Newton, who died in 1727. Now we want to step back in time to the ancient Aristotelian and modern Cartesian alternatives, both alive and well in 1650, and in this and subsequent chapters trace the process by which Newtonian science in all of its branches became dominant. The sciences that the Hebrew texts of the 1850s extolled owed their foundation to the concepts set forth nearly two centuries earlier in the *Principia*. Chapter 1 examines Newton's masterpiece; now let us look back at the alternatives available before 1687 when the book first appeared.

Let us suppose that some of the Hebrew readers were learning the scientific concepts for the first time, that they were in effect like European readers of the seventeenth century. What conceptual apparatus did they and their seventeenth-century predecessors have to discard in order to accept the new science that came to prominence with Galileo's 1633 confrontation with the Catholic Church? If they had received a classical education in Hebrew, Latin, and Greek, they would have revered nature as God's creation, like their Christian counterparts, particularly in Protestant Europe; and the new physics and its attendant science were presented with that religious tradition in mind. Nothing would have had to be discarded from such a religious orientation. These readers might have been familiar with atoms and various machines known to Archimedes from classical, pre-Christian texts. It is interesting for the larger story of paradigm conflicts in the West that the Hebrew textbooks cut out everything from ancient Alexandria to Galileo: "For approximately 1800 years there was no important discovery relevant to mechanics until the year . . . 1603 [when] the sage Galileo discovered three important laws."[4]

Would that the history of science had been so simple. The experimental and mathematical model associated with Galileo, rendered mechanical by Descartes and perfected by Newton, seemed the obvious way to do science only decades after the founders had slowly put it in place. For at least two generations, from the 1630s to the 1690s, competing scientific languages, scholastic or Aristotelian, Cartesian, alchemical, and Newtonian, resulted in a babble of tongues that could sometimes be spoken almost simultaneously by the same person. In addition, there remained the dogged opposition of the Church to the notion that the earth moved. Heliocentricity was to be treated only as an hypothesis and not as the reality of the sun resting and the earth moving. General piety about nature as God's creation still stood; but in general in Catholic Europe, the Bible could only be interpreted as putting the earth in the center with the sun moving around it.

For centuries in the West, until the time of Galileo, the privileged philosophical approach to nature owed its origins to Aristotle, as reworked by the medieval scholastics. It was well into the seventeenth century before Galileo's contemporary, the French philosopher Descartes, offered a comprehensive alternative to this version of nature. He insisted that all motion occurred because of contact between bodies and that clear thinking aided by mathematics gave the study of nature a new future. Within a generation after Descartes's death in 1650, yet a third understanding of nature came from Newton and the Newtonians. They laid far greater emphasis on experimentation, and Newton did not shrink from asserting that invisible forces, and not the push and pull of bodies, caused movement in the heavens.

Both the Cartesian and Newtonian models faced major obstacles. Since the Middle Ages, the Catholic Church had embraced a Christianized version of Aristotle and taught it in every school under its control. Many Christian devotees of the new science compartmentalized their minds and simply observed that scripture and nature said different things and both required admiration.[5] Most such thinkers knew little about Jewish learning, so we can understand why, by the mid-nineteenth century, a Hebrew textbook could repay the compliment and block out the nearly eighteen hundred years of largely Christian and scholastic thinking about nature that preceded Galileo. Hebrew scholars may also have known little about the Church's favored form of philosophical learning, scholasticism.

As taught by the schoolmen known as scholastics, Aristotle divided nature into matter and forms and approached the first by describing the second. Matter possessed no underlying conceptual unity such as it received from notions like the atom, force, gravity, and so on. Similarly, motion resulted from tendencies implanted in the form. It was in the nature of heavy bodies to fall downward to the heaviest of all bodies, namely the earth. Scholastic textbooks began with types or forms of matter; by contrast, modern physics and chemistry began with basic definitions of atomic structure and forces.

It took over a hundred years, from roughly the 1650s to the

1750s, to dethrone Aristotle—or more precisely, to remove scholasticism from every nook and cranny of the European educational system. By the 1650s scholasticism had become an obstacle to the acceptance of the new atomic science. But that negative assessment should not obscure its positive contributions, and the debt to Aristotle's role in the development of modern science must be acknowledged. His *Physics* promised that it was within the human capacity to know "the principles, causes, or elements" of nature and that there was "a natural way of doing this" by moving "from what is more obscure by nature, and clearer to us, towards what is more clear and more knowable by nature."[6] Aristotelianism possessed remarkable resilience. Many of its practitioners wanted to make natural philosophy autonomous; and, although always tied to Catholic doctrine, Aristotle's physics stood as its own discipline within the curricula of the universities. Some scholastics also took in new influences, the revival of neo-stoicism and hermeticism in the sixteenth century being but two examples.[7] The neo-stoics wanted the scholastics to think more about human nature and the nature of the state while the hermeticists, strongly attracted to alchemy, wanted to reform medicine. It was as if Aristotle provided a mantle under which various branches of learning developed and referenced the tools he provided, particularly logic and rhetoric.

By the thirteenth century, Thomas Aquinas had purged Aristotle of his pagan beliefs. God was defined by the scholastics, followers of Aquinas, as the creator of the universe, which was not eternal as Aristotle had asserted. By the fifteenth century, the mathematics needed for accounting, mapmaking, and calendars had become commonplace. Alchemy and astrology were also widespread. Though both have now been defined as forms of magic, at that time alchemy helped create pharmacology while astrology aided doctors who cast horoscopes to intuit details of a person's life that might actually help in a diagnosis. All these practices could find a place within the scholastic vision. Matter was to be approached through its qualities, its appearances, and shapes; and it was to be imagined as possessing inherent tendencies to rise or fall. Numbers could record debts and credits, but they could also be seen to possess mystical significance.

As late as 1687, the year of the *Principia*, an author making an alphabetical list of all the known branches of learning put astrology after arithmetic.[8] At that time (despite the title of Aristotle's great masterpiece), there was still nothing that could have been called physics as we know it (in English, "physic" meant medicine). The many aspects of that discipline were separate arts, the closest to modern physics being what was then called mechanics. In Paris the new Academy of Sciences devoted one day a week to physics. In the late 1660s and 1670s, this time was largely spent applying alchemical concepts to the distillation of vegetables and plants as a way to establish their essences and medical usefulness.[9] The leader of the effort, Samuel Duclos, also wrote about the transmutation of base metals into gold, the holy grail sought by all believing alchemists. Within an Aristotelian framework, mechanics "consider[ed] the quantity of moving forces . . . the gravity of a body is a certain capacity of falling downwards."[10] The richness of scholastic approaches to nature gave a wide berth to a variety of medical, engineering, and artisanal practices—the arts and sciences flourished in tandem. They would begin to diverge once a new and revolutionary matter theory—what we call the atomic and mechanical—came to prevail.

Galileo, Descartes, and Newton each contributed mightily to the revolution that made matter mechanical, that is, atomic, unmoved unless acted upon by other matter, and best understood because capable of mathematical and experimental explication. In the new understanding of nature, much of what the scholastics taught was discarded—qualities, forms, tendencies, the incorporeality of the heavens, the stationary earth—and replaced forever by atoms and forces and the uniformity of terrestrial and celestial matter. By 1650 in England, France, The Netherlands, and Italy, clusters of natural philosophers had become convinced that the way forward lay in the new conceptual structure known as the mechanical philosophy, best described by Descartes. Matter in motion, the "spring" of the air as it was pumped out of sealed glass containers, the weight of water and the speed of its descent compared to the fall of other bodies became the stuff of experiments and calculations. By the 1670s it had become possible to think of light as a body with weight and hence with a finite speed. The Danish astronomer Olaüs Roemer, working in L'Académie royale des sciences in Paris and sharing space with the alchemists, made that discovery.[11]

By the time of Newton's death in 1727, alchemy (in which he believed) and astrology (in which his great rival, the German philosopher Leibniz, believed) had begun to appear quaint. Still very popular, the magical arts no longer commanded the attention of the trained, practicing experimenter. They required unique insight that was difficult, if not impossible, to replicate by experiments. Yet alchemy remained popular among the less learned well into the mid-eighteenth century, especially in France and Germany.[12] In the Dutch republic during Newton's lifetime, the committed mechanist Christiaan Huygens ground lenses for microscopes and telescopes so as to better see nature. This was the same Huygens who in Paris worked in the library of the king, discussed the working of cannons, and shared space with the alchemists also active in the first decade or more of the new Academy. Across the Channel, Robert Hooke lectured on earthquakes as if they were natural phenomena, and not primarily signs of divine wrath.[13] All mechanists assumed that the matter being studied possessed the same atomic structure and all phenomena obeyed a set of natural laws, most still largely unknown. Invisible forces abounded in nature—hence the intense interest in electricity and the nature of light; but the way forward toward understanding them lay in experimentation and demonstration, not in magical claims to special mental powers and secret recipes. In 1728 the maker of one of the first encyclopedias, Ephraim Chambers, remarked that at the time there were still three schools of natural philosophy: the Cartesians, the "peripateticks (that is, the Aristotelians), and the Newtonians."[14] Within a dozen or so years, at least in Britain, as the paradigm decisively shifted, the first two largely disappeared. There the alchemists had first absented themselves from the public view after 1660. Indeed the English civil wars of the 1640s and the radicalism of the 1650s had tainted the magical arts with threat of social upheaval. By the 1660s at the Royal Society of London, no one talked openly about alchemy while across the Channel in Paris, no such stigma had yet blighted the future of alchemical work. Only in the 1680s did alchemical language disappear from the proceedings of the Academy of Science, gradually replaced by the mechanical and the Cartesian.

In France Newtonian assumptions and methods took longer to achieve dominance. Take the Royal College in the beautiful cathedral town of Chartres, today about an hour's train ride from Paris. It was old and respected, and in the 1730s the college had a priest, M. Morin, as the professor of philosophy. His Cartesian textbook tells us a great deal about the cultural posture of the new science in early eighteenth-century France. Boldly he called the book *Mécanisme universel,* and he advertised on the title page that it was based on observations and experiences made at L'Académie royale des sciences in Paris and the Royal Society in London, as well as his own inventions. Lectures, he proudly noted, had also been given before the Archbishop and "all the town." By the 1730s, throughout Europe public lecturing in science had begun in earnest.

Morin put his hand on machines and used them to explicate physical principles, including the vacuum, which was first shown to exist by Robert Boyle using an air pump and a bell jar. At the same time Morin began by explaining the usefulness of scientific work for theology and the value to be placed on the logic exercises so beloved by the followers of Aristotle.[15] After the flattery, Morin

damned the scholastics by faint praise: "If our brothers would investigate nature, not by authority or imagination, but by reason
and experience, they will advance [physics] mightily in the knowledge of the truth." Aristotle, he went on to say, proved things *out of
philosophy,* but Morin proudly proclaimed experimental demonstration as the way forward. He invoked the spirit of Robert Boyle,
the founder of the experimental method and Newton's friend.

Boyle can be said to be the inventor of experimental techniques
that he explicated in such a way that they could be copied. By creating a vacuum with his air pump, Boyle disproved Aristotle's claim
that nature abhorred emptiness. Morin used Boyle to attack sorcer-
ers, magicians, and diviners by associating them with the devil.
Cautiously, he endorsed Copernicus's claims but only as a "proba-
ble" hypothesis, a distinction that the Church required. Newton's
discussion of universal gravitation also received a mention in this
early French textbook on mechanical philosophy; but the existence
of gravity was treated as only an hypothesis regarding how the
heavens move, and not as good an explanation as Descartes's contact action. Newton had offered mathematical proof for the action
of universal gravitation, but he had asserted no mechanical hypothesis to explain how it worked. Next Morin offered a lengthy treatise on the human body seen as a mechanism. He introduced a
theme that would remain constant in the early assimilation of science: It must be relevant to the ways we understand the human
body and medicine, although how relevant was by no means clear.
Finally, elaborate attention was paid to instruments, magnets, and
pumps.

Morin's book passed the French censors, which was no mean
feat. The Jesuits opposed the science of both Descartes and Newton
and they were powerful. As a foremost historian of French higher
education in the period puts it, "If Newton finally triumphed in
France it was probably over the corpse of the Jesuit Order."[16] The
Jesuits only accepted Newton in the 1750s. Morin perfectly re-

flected the wide scope of academic science in France by the 1730s, just as his book displayed the eclecticism needed to soothe the Church, which still espoused Aristotelianism, address Newtonian science, and still hold on to a Cartesian framework. Anyone who wanted to champion Newton to the detriment of the scholastics and the Cartesians had to have something novel to say, and Morin tried to be a state-of-the-art mechanist while still keeping his Catholic critics at bay. As the paradigm shifted away from Aristotle, and finally away from Descartes, resting then on Newtonian science, the French followed the British and the Dutch as exponents of universal gravitation and everything that it implied.

The French Newtonians

Interestingly enough, by the 1730s only Newton's law of universal gravitation remained among the laws for which no physical or experimental demonstration had yet been provided. Aside from his brilliant mathematics, Newton relied on analogies to local motion, weights, pendula, and so on. But how could one measure gravitation at work on the earth itself and then prove its physical existence? Ironically, the physical evidence that confirmed Newton's law came not from work done in England, but in France, by Pierre-Louis de Maupertuis. He rose in the learned circles in which the Parisian Académie royale des sciences reigned supreme. Getting the Académie's attention required originality as well as good fortune. Maupertuis had both; and on an expedition financed by the king of France, he got a golden opportunity.

As Morin's textbook demonstrated, pockets of resistance existed all over learned France to the acceptance of Newton's principles. The naysayers who followed Aristotle demanded a vacuum. Descartes had insisted on a plenum, on a universe filled with a fine matter or aether as well as the large bodies of the stars and planets. The Newtonians assumed action at a distance in a void with gravity operating from planets and the sun through the emptiness of space. Descartes, and indeed all Continental mechanists, required contact action between bodies for motion to occur. Maupertuis recognized that demonstrating the shape of the earth would be "of very great importance," not only for navigation but also because that achievement would prove—or disprove—one of the key postulates of Newtonian science. If the earth flattened at the poles, Newton argued, it would be because of the centrifugal force at work at the earth's center; this force was strongest at the equator, where the earth experienced the "crushing" effects of increased gravitational forces at its exterior. But how to prove such an hypothesis? To make his mark on the science of his day, Maupertuis had to get very cold.

Journeying toward the north pole in the wastes of Finland, Maupertuis used an elaborate pendulum device to show that, when controlled for temperature changes, a pendulum swung more quickly near the pole. He used pendula of different substances, each with a different specific gravity, and the same quickening occurred. Maupertuis also used a telescope to measure a flattened arc of the earth. Triumphantly, warmer and back in Paris, Maupertuis announced that "all the experiments which the Academicians, sent by the King . . . conspire with ours, to make the increase of gravitation towards the pole, greater than according to Sir Isaac Newton's Table, and by consequence the earth flatter than he has made it." Maupertuis claimed that he even outdid Newton in the accuracy of his measurements. In addition the increase possesses uniformity, "the Gravitation increases toward the Pole as the Square of the Sine of the Latitude."[17] With a single expedition, Maupertuis had brought innovative science to Paris and tied it to the star of Newtonian natural philosophy.

At the same time Maupertuis demonstrated the importance of institutional and royal sponsorship. "If I might be allowed to do justice to the courage and talents of the rest of my companions, it would appear that the work we were engaged in, difficult as it was, must become easy in such company and with such assistance."[18] In 1737 before a meeting of L'Académie royale des sciences, Maupertuis announced his findings and set in motion what would become nearly a century of brilliant Newtonian physics that would call Paris its home. In the process he added more evidence that served to defeat the Cartesian model of the universe. Maupertuis went on to pioneer the application of Newtonian ideas and sophisticated mathematics to the study of the heavens—or geodesy, as it was called. As the leading historian of his work, Mary Terrall, puts it, "[Maupertuis's] commitment to the natural philosophy of gravity cannot be separated either from his desire to spearhead a distinctively French Newtonian physics, of which geodesy would be a part, or from his personal ambition."[19] By 1750 in France Morin's textbook would be obsolete as new reputations were made and Newtonianism gradually triumphed. The first clear breakthrough came with the appointment in 1740 of Pierre Sigorgne to the chair of philosophy at the College du Plessis in Paris.

The Formal Institutions of Science

Maupertuis's success also illustrates the power of scientific societies and academies to promote and underwrite learning. Indeed we now take it for granted that the appearance of the new science in Europe from the time of Galileo onward implied the formation of clubs, societies, or academies. They turned up whenever and wherever interest in natural philosophy surfaced, and they played a decisive role in turning the attention of the learned away from Aristotle, toward Descartes and finally Newton. The Lincei in Florence, L'Académie royale des sciences in Paris, the Royal Society in London, and the Academy of Sciences in St. Petersburg provide dates when interest in the new science took hold: in 1610 in Florence, in the 1660s in Paris and London, and in 1725 in Russia. The fluidity of borders, between the mechanist and the anatomist, between the optician and the geometer, required social interaction and so too did experimental demonstration. One of the first acts of the St. Petersburg society was to send out letters in Latin to all the other scientific societies in Europe. As a result, a series of communications and collaborations began between St. Petersburg and the Royal Society in London. The same letter sent to Sweden—just recently an enemy of Russia—produced a good response from its university in Uppsala.

Membership in, or even just association with, a scientific society also suggested, if not confirmed, a minimum competence in matters both technical and scientific. In 1731 the French Department of the Marine got a report on the work of an English company hired to drain a mine in Spain. In the same decade the English engineer Richard Newsham tried to find employment with the same French ministry, and he sent a certificate from Jean Desaguliers, official experimenter for the Royal Society, who attested to his training and abilities. Newsham's pumps were so good, Desaguliers wrote, that they were used to bring water into the house of the Royal Society. Around the same time, yet another English engineer recommended himself as approved by the Society and as being skilled in mathematics and mechanics.[20] The prestige of British mechanical science was beginning to be recognized, and having an affiliation with a formal scientific society aided the rise to prominence of any career.

By the 1730s the interests of the Royal Society had turned resolutely toward the physical, mechanical, and mathematical, and Fellows interested primarily in natural history (and, by implication, agriculture and the land) were not being elected to positions of leadership.[21] This interest in mechanisms augured the shift that occurred within British society after 1760 as more and more entrepreneurs become interested in mechanical devices and ultimately in power technology. Yet we should not make the Royal Society seem too business oriented. In the 1790s it regularly got reports from William Herschel, the astronomer supported by the King of England, about the possibility of life on the moon and even on the sun.[22]

By the last decade in the century, the Royal Society's primacy was being challenged by other coteries of scientific interest, including the Lunar Society in Birmingham and the Pneumatic Institution in Bristol. Unlike the conservative stance of the Royal Society, both groups had radical political associations and, in their way, were at the cutting edge of scientific thought and application. In Birmingham just about every leader in the new technology of steam belonged to the Lunar Society—named for its habit of monthly meetings on the evening of the full moon. In Bristol in the 1790s, the new science of pneumatics, or gases, which was pioneered by Thomas Beddoes, promised to offer cures for everything from consumption to depression. These cures never materialized, but in chemistry Bristol was the place to be. Humphry Davy, the finest chemist of his age, got his start there, and he moved to London in 1801 to lecture at the Royal Institution. It too became another competitor to the primacy of the Royal Society.

In Paris advancement in the ranks of l'Académie required considerable skill, as the young Maupertuis would have been the first to say. Yet throughout the eighteenth century, its members—including Jean d'Alembert, Laplace, Lavoisier, and Chaptal—were by far the most distinguished mathematicians, chemists, and natural philosophers of the age. The form of its gatherings shocked English observers. Members sat eating at a round table and everyone talked at once. In London the Royal Society's meeting room resembled a modern lecture hall and members were expected to listen politely and ask questions. The Académie's round table was for equals—that is, most members were aristocrats or, at the very least, recipients of the largesse of the king. In contrast, the London lecture hall of the Royal Society attracted men of vastly diverse backgrounds whose formality made affability easier. Both societies maintained intermittent contact from the 1730s onward, and an exchange of weights and measures sealed the acquaintanceship. Yet unlike the Royal Society in London, L'Académie royale des sciences had only very limited contact with the other French academies and societies in the provinces.

The social nature of science meant that when the paradigm began to shift—as happened in France by the late 1730s—group cohesion tended to solidify the transformation. If science had just been a matter of books read quietly in the study or solitary experiments in the kitchen, lone enquirers might have remained Aristotelian or Cartesian for many more decades. The social and institutional nature of scientific enquiry may make a paradigm hold long after its usefulness is in doubt; but it also means that when a house of cards crumbles, the socially reinforced tendency is to jump on board and embrace the new understanding.

But the social by its very nature also entailed a babble of less expert tongues. In both London and Paris, there were men in the societies who made no contribution of any consequence to the study of nature. They were curious, and they enjoyed the status that membership conferred. The vision and purview of such diverse social groups may be rightly described as provincial and imperial as well as eclectic. In the same meeting L'Académie royale des sciences discussed the geographies of Turkey, China, and Armenia and tigers in China. Then a Cartesian priest, Father Malebranche, contributed a paper on light and colors and the generation of fire while another gentleman described a new sort of floodgate to help control navigation on the Seine. Still another contributor said that he had found "a commodious way of making use of fire to move machines."[23] In the 1750s the opening proceedings of the Haarlem society in the Dutch Republic offered discussions of logarithms, bones, fungus in the uterus, finding one's way at sea, electricity, the planet Mercury, human free will, and eclipses of the sun. In addition, the Dutch society published a treatise on inoculation arguing that there was nothing about the practice that violated religion.[24] Of course, windmills also figured in the discussions of the new society.

Utility was never far from the agenda of the European societies. In this respect the Royal Society probably led the way. As early as the 1680s, it was discussing the use of machines to save on labor costs. At the time no patent would have been granted to a device that was intended to save labor because nearly 20 percent of the population had no visible means of employment.[25] Some of the earliest experiments with steam engines were conducted under the auspices of the Society. Later in the eighteenth century, its president, Joseph Banks, avidly invested his personal fortune in canals, mining, and other projects with an industrial focus.

For much of the eighteenth century, the definitions of nature employed in London and Paris were so commodious as to embrace folklore gained from travel along with experience and experiment. The mating habits of insects appeared in the minutes along with endless discussions of human anatomy. Technology, from the microscope to the barometer, was presented for observation. So too in Paris a report arrived about how "European women who go to Batavia [a Dutch colony] cannot suckle their children, their milk being so salt that they will not take it; whereas, the milk of the Negresses, though their diet is the same, is sweet and pleasant as usual."[26] The sense of all the world being under European purview probably enhanced the cosmopolitan aspect of fraternizing around science while simultaneously rendering the rest of the world exotic and effortlessly exploitable. As was typical of the major European colonies, both the Dutch island of Batavia and the French colony of Saint Domingue (today Haiti) possessed scientific societies.

In none of these settings were women expected, or even allowed, to be members. Sometimes they observed public lectures or took private courses in science, which were available in any major American or European town. One exception to the gender rule surfaced late in the century. Formally established by and for women in the town of Middelburg on the southern Dutch island of Walcheren in the province of Zeeland, the Natuurkundig Genootschap der Dames (the Women's Society for Natural Knowledge) met from 1785 to 1881, finally closing its doors in 1887.[27] This society challenges our stereotypes of women, the physical sciences of the day, and the intellectual interests open to women in the early European republics. Its overall membership of approximately two hundred women included the elite of the society as well as the wives of local clergymen. The women studied the standard textbooks of the era and bought scientific instruments, which they used both at their meetings and sometimes in their homes. The Natuurkundig Genootschap der Dames illustrates vividly the integration of science into the fabric of domestic life among the highly literate, a process at work throughout the Euro-American world.

As noted in Chapter 1, since the time of Robert Boyle and the Boyle Lectures of the 1690s in England, and Bernard Nieuwentyt in the Dutch Republic, a Protestant sensibility had linked natural philosophy to theology. A vast sermon literature offered God's work to the educated laity as the way of grasping the wisdom of the creator.[28] Even if this physico-theological rationale for science was not yet generally accepted in the setting of Middelburg—opponents still forced scientifically interested ministers to defend the compatibility of science and divine revelation in public[29]—it was strong enough to inform the work of the Middelburg women. They sustained the only scientific society established for women and financed by them—at least so far all the evidence points to their primacy.

Despite the gender barriers, we should not imagine that watertight compartments separated men and women with an interest in science. In Philadelphia the widow of the schoolmaster William Johnson, who had been made a correspondent of the society that would, in 1769, become the American Philosophical Society, donated his natural history collection to that group. True to the passion for electrical experimentation that was common in the American colonies, Johnson had made a living lecturing to both men and women on "the nature and properties of electrical fire."[30] He was typical of the men who made up the coterie for science in the colony of Pennsylvania. Women can also be found in the audiences of scientific lectures in provincial Britain and The Netherlands. Yet overwhelmingly science became the work of men.

In Philadelphia Benjamin Franklin obviously stood as the most renowned and innovative member (and a founder) of its local society, but typically shopkeepers, clergymen, merchants, artisans, and small farmers filled the chairs. The prospects for such a society, "considering the infancy of your colony[,]" was judged to be dim by naturalists in London. But John Bartram, to whom the pessimism had been conveyed, went along undeterred. Bartram's botanical investigations made him famous by the 1740s and, along with Benjamin Franklin, he set out to imitate the Royal Society and the Dublin Society and to establish an American equivalent. It took them over twenty years to collect the necessary critical mass of gentlemen with interests in natural knowledge.[31] Again, the overarching natural philosophy dominant in Philadelphia became Newtonian, but there too we also find an eclectic range of observation and collecting.

Once established, the American Philosophical Society gave European philosophers like the Marquis de Condorcet a place to write with inquiries about the new world. They asked about everything from the behavior of mercury in a barometer to whether "black children born free and educated as such, have retained the genius and character of the Negroes, or have contracted the character of Europeans."[32] The literary and philosophical range of the American Philosophical Society resembles that of the many "lit-phils" in Britain. To this day, it admits members from every branch of learning. As the eulogist said at Franklin's funeral, his had been a "practical philosophy of doing good." Not surprisingly, a number of medical doctors formed the rank and file of the society, and subjects close to their interests were regularly discussed at the meetings. A similar pattern of interests can be seen at the meetings of the Royal Society and the Dutch Society in Haarlem. The divisions and specializations that lie at the heart of modern science, and that make medical training an entirely different set of curricula, were not firmly in place in Europe or America much before 1850. All of those interests could be found represented to greater or lesser degrees in academies for letters and science that, by 1800, extended from Boston to Bologna, Bordeaux to Brussels.

Other social changes common to the century were also prefigured in scientific circles. Inadvertently, science helped to invent and strengthen cosmopolitan social mores. National borders were crossed, as were social classes—within limits—because specialized knowledge was constantly being conveyed to those slightly less expert than the conveyor. There was nothing *necessarily* cosmopolitan and international about science—indeed, national rivalries and social nastiness were commonplace in scientific circles—but the practices of science, more than any other single new cultural phenomenon of the early modern era, more than reading or coffee housing, constantly threw male strangers, and some female ones, into sustained social contact. The social nature of early science arose in part because so few people in fact did it. Specialists branched out, and bringing others with them required the observation of experiments that bridged gaps in knowledge. Those different levels of expertise or interest could be accommodated by seeing and touching phenomena. The faith developed, and by 1800 became unshakeable: Through the linkage between science and philosophy, Europe could transform the entire world into its image and likeness.[33]

In Russia during the 1720s, Tsar Peter the Great sought to copy the behavior he had observed in Paris and London. Again the goal was imperial and imitative of the power enjoyed by the western European states. In Peter we have once again found our anthropologist. We can use his eyes to help explain why the social configuration of the state-sponsored or small, private society seems so integral to the growth of Western science. He saw such academies as "nothing if not a society (gathering) of persons who assist each other for the purpose of the carrying out of the sciences," and then tellingly said that it was essential to verify experiments in the presence of all members because "in some experiments many times one demands a complete demonstration from another, as, for example, an anatomist of the mechanic, etc."[34] The diverse interests of scientific practitioners also matched the diversity of occupations and backgrounds among members of the various scientific clubs. Matched only by the Masonic lodges in some parts of Europe, scientific societies were places where Christians and Jews mingled freely. In fact, nine Fellows of the Royal Society in London were Jewish.[35] Scientific societies strengthened the bonds of civil society and created zones in which relative freedom of communication became possible. Their cosmopolitanism acted as a natural check on the ambitions of absolute monarchs and closed elites to keep power and learning solely for themselves.

In the 1980s, during one of the tense times of the Cold War, our histories of science tended to emphasize the closed and sequestered nature of early science, seeing, for example, in Boyle's air pump a massively expensive device, an early form of Big Science, that precluded just about anyone—except someone as wealthy as Robert Boyle—from ever owning or using it. Here we wish to emphasize the openings and possibilities that formal and institutional interaction around natural philosophy offered. To be sure, all the official academies were elite gatherings and patronage was ever-present. Even the Royal Society of London and, in Haarlem, the Hollandse Maatschappij der Wetenschappen (the Dutch Society for Sciences), both private and dues-paying, required a stiff entrance fee and high literacy. But going to such gatherings or just reading their proceedings, which were widely circulated and translated into many languages, opened the mind.[36] Oftentimes one person constituted the link between two societies in different countries. In the case of Bologna and London, Thomas Dereham, a Fellow of the Royal Society, provided the contact long before the two societies had formal relations. By contrast, however, the links between the Berlin academy and the Royal Society in London were weak and only took shape in the second half of the century.[37] By then, the Berlin academy had been reorganized and renamed by Frederick the Great. He was smart enough to offer Maupertuis the presidency of the refurbished society, thereby in a stroke giving it considerable international caché.

Science and State Building

Strikingly, the early scientific societies, or the theorists who preceded their establishment, advertised their advantages to emergent national states throughout the seventeenth century. In the Elizabethan era, in the aftermath of English success against the Spanish Empire, Sir Hugh Plat argued that joining natural philosophy to the aims of the state was essential for the good of the nation. Taken up by Francis Bacon, the argument took hold and almost every European state thought that science offered unimagined benefits. During the period of the English civil wars, when the Stuart king, Charles I, was chased from the throne and then executed by Cromwellian radicals, the good of the Commonwealth was thought by some like Robert Boyle to be dependent on the growing exploitation of natural knowledge.[38] This notion informed debates about inventions and investment, about patents and stocks—on the one hand, suggesting that the discoveries of philosophers could lead to great wealth while, on the other, raising alarm about those involved in false promotions, in what then came to be described as "bubbles." In the English circumstance, these arguments complemented the strategic objectives of a nation increasingly at odds with Continental rivals.

As the Elizabethan merchant Sir Thomas Gresham determined, it was essential to develop the necessary skills and knowledge to be able to compete with the Continental countries at least in matters of commerce. For this reason, Gresham left funds in his will for the development of teaching and for public lectures accessible to tradesmen and artisans who would, otherwise, have no entrée to the learning of the universities. Gresham's vision made a great deal of sense for a nation surrounded by the sea, where advances in seamanship and skill in navigation were as crucial to national survival as commercial success. In England, men of trade and manufacture increasingly mattered to the strategy of a relatively small nation state. The same Gresham who promoted science built the Royal Exchange in the heart of London where traders in stocks and goods from all over the world gathered.

Even under the Commonwealth, before the Restoration of the Stuart monarchy in 1660, scientific and economic concerns were sometimes connected.[39] Indeed, itinerant lecturers could be found in the 1650s traveling the English countryside, discoursing on such matters as weapons, chemistry, and the furnaces necessary for the smelting of metals. Again Bacon's philosophy offered the best articulation of such a national strategy. It was also adopted by the Royal Society when, soon after its creation in the early years of the Stuart Restoration, it came under attack by those suspicious of its activities in an age that was highly sensitive to whiffs of religious deviance. The Society claimed it promoted the trades by gathering information about them, which could then be spread to others interested in the utility of natural knowledge. This was not a novel idea. It had predecessors on the Continent, most notably in the Netherlands and in France. In the new Royal Society, a program was established to gather a "History of Trades." This, in the end, failed to produce much worthy of dissemination because of the resistance of craftsmen to reveal their secrets and their desire to protect their livelihood. There was also great resistance in the Society, instigated by Fellows who sought to preserve its social exclusivity, to concerting with "private men" of the trades who sought wealth above pure knowledge.[40]

The Royal Society, since its earliest days, had nevertheless presented itself as absorbing learning from those with skill, as much as from those with philosophical curiosity. The utilization of nature was a theme that its early promoters tried to exploit, especially as it could be linked to the strength of the state. Whatever might have been intended, the Society was not in fact formed "of all sorts of men, of the Gown, of the Sword, of the Shop, of the Field, of the Court, of the Sea; all mutually assisting each other."[41] If this was a program for the recruitment of useful Fellows, then it was a dismal failure. Increasingly, to be a gentleman of an enlarged curiosity was sufficient to be recommended to the Society. It did not hurt if one were also well connected. This was, after all, a patronage culture. The result was not only the collapse of its History of Trades, but also a Society that by the 1690s appeared to some to be languishing, without direction, if not in a state of decay.

Absolutism ended in England in 1689 and so too did the dream that the Royal Society harnessed to the interests of the monarchical state would engineer the power of both. In capturing the power of science when institutionalized, the Continental absolutist states of the eighteenth century, with their cash subsidies of the scientific societies, led the way. In one area, that of mining, the German princely states were critically important in creating engineering corps with good technical and scientific training. The Russians and the Swedes also gave a professional and scientific education to their mining engineers. Such schools became the model and the mining schools were imitated in France, Italy, and Spanish-speaking Europe.

In one absolutist state after the other, the establishment of an academy of science or an engineering school signaled that, even in Catholic Europe, the new science had to be embraced. In 1779 the queen of Portugal set up such an academy as well as a school for the professional training of military engineers.[42] In Spain during the reign of Charles III (1759–1788), new scientific institutions were created under royal patronage and the rhetoric of utility and technical progress became their justification. Yet well into the nineteenth century, science remained controversial in Spain and many questioned its role in relation to true religion and even its usefulness.[43]

Generally speaking, Protestant Europe and Catholic France fostered institutionalized science more easily and with less friction over religion. In Prussia Frederick the Great made the bold move of appointing Maupertuis in 1740 to head his academy because, relative to France and Britain, Prussian science and society were underdeveloped. The guiding light of the previous generation, Gottfried Wilhelm Leibniz, had tried, without great success, to spread a dedication to science throughout the whole of German-speaking Europe, roughly the territory known as the Holy Roman Empire. He wanted scientific societies everywhere that would—in top-down fashion—foster industry and invent economic progress.[44] Frederick the Great tried to start where Leibniz had left off and his academy in Berlin became respectable, although in no way as dynamic as its Parisian counterpart. Nothing quite matched the brilliance found at the French academy from Maupertuis well into the 1790s.

With a few exceptions, the universities of Europe lagged behind the academies and societies as places that made scientific enquiry happen. In Germany the universities of Halle and Göttingen were more advanced in the institutionalized teaching of science than almost any other university in Continental Europe.[45] In Britain in matters scientific, only Cambridge and Edinburgh were their equals, if not their superiors. At Cambridge, professors in the eighteenth century emphasized utility and the promotion of chemistry in industry. In general the British universities of the eighteenth and early nineteenth centuries played little or no role in the emerging zeal for mechanized industry and coal mining. These initiatives would be left to mechanically savvy entrepreneurs whose education in machines and mechanical principles had been real, but often informal.

On the Continent, and particularly in Germany, the story was different. To be sure, Halle and Göttingen were exceptions, and not the rule, in the German-speaking lands. Yet in them, the state as sponsor of scientific inquiry, and especially of its application, finally took hold. In the nineteenth century, the German state gave vital support to the chemical industry; and by the 1860s, assisted by the universities, chemical science had propelled Germany into the forefront of Western science.[46] But that achievement was slow in coming, and in 1800 no one would have predicted that Germany, and not France, would be the Continental powerhouse in the application of science to industry. At the same moment no one could imagine catching up with the British in the application of mechanical knowledge to the manufacturing process. That level of mastery had been achieved without a direct partnership between the Royal Society and the state. In Britain private men possessed of money and access to an increasingly public science had made the nascent Industrial Revolution happen. They were not, however, without informal resources that reinforced their reliance on scientific learning.

Informal Clubs and Publishing

Science seemed to require clubbing; and although rivalry between experimenters was common and often vicious, a modicum of politeness kept any club afloat. Protestant Europe proved exceptionally receptive to Newton's science and some of the earliest, informal fraternizing in the new polite and cosmopolitan mode occurred around it. In places as small as Spalding in Lincolnshire, a literary and philosophical society began to meet around 1710; and, within a decade, over 300 men were members in a town with about 500 families. On the Continent The Hague, a town of some 35,000, also supported a loose coterie of early Newtonian advocates. There one of the earliest groups that sought to disseminate Newton's science set up a secret club and into it came the young Dutchman Willem s'Gravesande.[47]

The little society in The Hague where members were Protestants and called one another "brother" also published a journal in French that disseminated Newtonian science far and wide. The minutes reveal the group to have been jovial, even risqué, and an attendant society with many of the same members left a meeting record of its drinking exploits. In those records the handwriting deteriorates almost by the sentence.[48] In one or another of these two societies, we find a postmaster in Brussels with literary interests; French Huguenot refugees; a German bookseller; the chaplain to an English aristocratic lady resident in The Hague; the young Willem s'Gravesande, who led the University of Leiden to eminence in Newtonian science; and the journalists who made up "the corps of the society of authors who compose the *Journal Litteraire* [sic]."[49] It was arguably the first journal on the Continent to disseminate Newtonian science. The interests of the members and their friends branched out into publishing more generally and also to the circulation of clandestine literature, much of it hostile to all religion. Some of the characteristics of the group in The Hague have a decidedly Masonic look, which is not surprising, given that they were in close touch with English intellectual life. As we saw in Chapter 1, freemasonry began in England and Scotland.

The relative absence of specialization in such clubs might have led to a frantic search to divide up areas of enquiry so as to conquer, to create in effect a narrow specialization with clubs solely for botany, or mechanics, or mathematics. But curiously the eighteenth-century impulse among the learned, both social and intellectual, was to unify, collect, combine—in short, to invent the encyclopedia.

The zeal to classify appears most obviously and, in the first instance, not among the writers of books but among their sellers. The large publishing houses and book dealers provide some of the earliest evidence we have for the invention of new classificatory schemes. By 1700 the sheer volume of new books and range of their topics seems to have forced upon them unprecedented attempts to classify and order. Almost simultaneously, the same circles that sought to classify came to include men with significant interests in natural philosophy and its dissemination. On the Continent Prosper Marchand, one of the editors of the *Journal Litteraire,* invented modern classification systems for his book business's inventory just as he disseminated Newton's science.

In England two of the earliest encyclopedists, John Harris and Ephraim Chambers, belonged firmly in the orbit of the Royal Society. Harris, an Anglican preacher, also gave a series of Boyle lectures, and he was a firm believer in physico-theology. He invented a two-volume technical lexicon, or "an universal English dictionary of arts and sciences" where under "jac," for example, later editions printed the following: "*Jack* in a lantern, a certain meteor or clammy vapour in the air . . . commonly haunts churchyards . . . jack in a ship is that flag which is hoisted . . . *Jacob's staff,* a mathematical instrument for taking heights and distances . . . *jactivus* . . . a Latin word, signifying in the law, him that loseth by default."[50] Throughout his dictionary Harris focused on integration, bringing the new science and its instrumentation into a universal alphabet that allowed "improper fractions" to reside between "imposthume" (a collection of matter or pus in any part of the body) and "impropriation" (a word for the profits of an ecclesiastical benefice being in the hands of a layman). Wherever possible, words were assigned meaning in the physical order of things: "Impulsive" refers to "the action of a body that impels or pushes another" while "incidence" refers to an optical angle that, in turn, includes a long digression on Newtonian optics with some background material on the work of earliest opticians.[51] The first edition in 1704 of Harris's text under the word "jupiter" never mentions the Greek god but devotes pages to the new astronomy. The word is squeezed between "Julian period" and "jurats," which "are in the nature of Aldermen, for Government of their several Corporations."[52] Each edition expanded ever more.

Harris's effort in the "j's" is no mere playing with words but rather a remarkable reordering of linguistic signs in the direction of the physical, material, and natural. "Impulsive" was no longer simply that which "drives or thrusts forward" as it would have been previously.[53] By the early eighteenth century, ordinary English dictionaries were beginning to include scientific terms, perhaps before dictionaries in any other Western language.[54] By the 1740s, a "society of gentlemen" had taken over Harris's dictionary after his death in 1719; they saw their enterprise as being in competition with Chambers's *Cyclopedia* of 1728, arguably the first modern encyclopedia.[55] Chambers's work became the prototype for the greatest encyclopedia of the century, produced in Paris by Diderot and over two hundred other writers.

In a primer on Newtonian astronomy done in the form of a dialogue between an aristocratic lady and a natural philosopher, Harris even implied that he had completed his lexicon with an eye to the education of women.[56] Such a target would have made good business sense, and Harris was attempting to make a living not only by his preaching but by his scientific publishing. Other publishers sized on the market of literate women and, as early as 1704, an enterprising schoolmaster invented the *Ladies' Diary, or Woman's Almanack*. Within a few years, with the engineer Henry Beighton as the editor, mathematical puzzles became the almanac's forte; and in 1720 the magazine presented problems requiring the use of Newtonian calculus.[57] Men came to dominate the proceedings of the journal, which lasted until 1840. Its great importance, like the women's scientific society in Middelburg, lies in the domestication of the new science, its entrance into home and hearth.

Economic historians credit the early encyclopedias with making a major contribution to the decline in access costs as knowledge could be acquired quickly through their alphabetical arrangement.[58] No one learned to make a steam engine from reading an encyclopedia, but it was possible to learn that such a thing existed and the basic principles used in its construction. In the small town of St. Hubert in the Austrian Netherlands (today Belgium), the local abbé wanted to exploit the metal deposits on his land and purchased a copy of Diderot's encyclopedia to learn how to construct wood-burning devices that would be strong enough to melt the ores. It is hard to know what, if anything, he got from Diderot's great work with its elaborate descriptions of all the trades, but clearly the priest thought it a good place to start.

The most ambitious as well as scientific and technical encyclopedia of the age came out of circles deep in the heart of the French Enlightenment. A consortium of publishers decided to bring out a French version of Chambers's *Cyclopedia*. Distinctively it paid attention to what Chambers called "artificial or technical" knowledge that complemented "natural and scientifical" knowledge. The latter encompassed all the branches of mathematics, including trigonometry, "physics and natural philosophy," ethics, religion, and theology. The French translators made short work of theology and religion; and where they did treat them, heresies were systematically introduced. In addition Chambers had roughly fifteen times the number of entries for the mechanical arts and manufacturing as Harris did. Each edition expanded those entries.[59]

The young Denis Diderot had spent time in the Bastille for publishing a pornographic novel and he needed the job that the publishers offered him. In every sense he expanded on Chambers's work, most notably by adding a set of magnificent engravings (some 2900 in all) to illustrate the range of artisanal crafts and by begging artisans to learn more from *savants*. Beginning with Harris and Chambers, expanded greatly by Diderot and his collaborators, the encyclopedia became a compendium that unified the sciences, rational and applied, with manufacturing and technical skill. The *Encyclopedia Britannica*, first published in 1788 to 1797, credited the "philosopher" as the source of all the principles needed by the architect, carpenter, and seaman. The encyclopedias reflected a very gradual transformation as scientific principles were incorporated into practical mechanics and that discipline, in turn, was applied to the manufacturing process. In retrospect by the 1820s, observers would say that what had happened, first in Britain and then throughout Western Europe, had been an Industrial Revolution. We now know that it had been aided significantly by the inculcation of scientific thinking combined with technical applications, and encyclopedias made for easy access to all those developments in applications.

General Education in Science and Its Uses

As noted earlier in this chapter, the teaching of Newtonian science slowly penetrated European and American lecture halls and universities. For example, Richard Bentley of Boyle lecture fame and later master at Trinity College, Cambridge, was instrumental in establishing professorships of astronomy and chemistry to be occupied by Newton's disciples. Most universities in Catholic countries were far slower in accepting the new science, first in its Cartesian and then in its Newtonian forms. And the approximately four hundred French colleges, which taught boys roughly fifteen to nineteen, missed an entire generation, compared to British and Dutch schools, by failing to teach Newtonian science until the 1750s. By contrast, Newtonian science was being taught in Dutch universities and academies by 1700 and universally there by the 1720s. Men and women in The Netherlands, as just about everywhere in Western Europe, could also take courses with paid instructors or tutors. In Amsterdam advanced lecture courses were given privately by Daniel Fahrenheit (of temperature fame), and he worked closely with the professor of physics, s'Gravesande, at Leiden. When Fahrenheit died in 1737, his estate advertised the sale of "his mechanical instruments used to demonstrate the Newtonian Philosophy."[60] He was as well connected in European natural philosophical circles as any university professor, and the surviving notes from his lectures show him beginning with Cartesian ideas and then switching over to Newtonian ones. Always he laid emphasis on application and usefulness.[61]

The support for Newtonian science must have been far stronger in most British schools and universities, but few records survive to prove that claim. Beyond Oxford and Cambridge, the evidence for what was being taught in the high schools—what the English call "grammar schools"—is almost entirely random before 1800. Incontrovertibly, the Dissenting academies, which taught boys around the same age as those in the French colleges, were advanced in the science of the day; and by 1750, they were favored by entrepreneurs as places to send their sons. The Watt family, made rich by their steam engine, would think of no other school in Britain except, of course, Glasgow, where the university was Presbyterian (hence Dissenting) and uniformly up to date in mathematics and science. In the eighteenth century, the same could be said for Edinburgh University and the acceptance and teaching of Newtonian science. By the 1790s, much of what got taught at all these British schools was closer to what we would call engineering than "pure," abstract, or rational science.

Gradually the applications of scientific knowledge, with a bias toward mechanics, became universally valued and taught. Whether in German, Flemish, or French, science penetrated curricula at every educational level. Prodded by a local society with technical and scientific interests in the 1770s, the Bishop of Liège in Belgium created a school for boys with an explicitly technical and industrial focus.[62] The area became one of the first to apply power technology to the extraction of coal in Continental Europe. In Lille in northern France, technical education also came early, at least by the 1790s if not the 1780s.[63] It too became an early industrial center. By the 1790s cheap handbooks could also be found in many languages. Aimed at artisans, they taught basic mechanics derived from Newtonian textbooks. In 1784–1785 in Philadelphia, Oliver Evans built the first automatic mill that moved on the principle of the overshot waterwheel. He did it based on his reading and then went on to put his knowledge into *The Young Millwrights' and Millers' Guide* (1795) so that other artisans might benefit as he had from self-taught mechanics.

Also in the 1790s under the impact of revolution, the French brought new men into government with new ideas for French education, particularly in the sciences. The French revolutionaries were convinced that their education lagged behind that in Britain and, being anticlerical, they knew whom to blame. In the wake of the French Revolution, new schools were established throughout the country, and at these *écoles centrales* the teaching of experimental physics was introduced. The explicit focus was on application. The principles of hydraulics were explained and so too were pumps. Steam engines were discussed and so too Newton's *Principia*. Remarkably the textbook chosen had first appeared in English in the 1740s and had been written by the first-generation Newtonian Jean Desaguliers.[64]

Teachers all over France wrote to the ministry of education to complain that they did not have the necessary equipment, "verbal descriptions are surely insufficient in the sciences where one can only instruct by a continual manipulation . . . I am doing everything I can to put knowledge into the hands of enlightened citizens capable of carrying the light to all the arts, to ameliorate the culture of the department and to establish manufacturing. . . ."[65] In Liége, now defined as a part of France, the professor of physics and chemistry in the new *école centrale* was so frustrated by the absence of instruments that he just quit. Demonstration devices were crucial. A few months later a small supply arrived; but three years into the new regime, there was still no chemistry laboratory.[66] By then small collections of mineralogical samples were being offered for sale throughout French territory by the Conseil des Mines. As described in Chapter 5, this uphill struggle of the French to enhance scientific education continued into the nineteenth century.

There is some evidence to suggest that the negative assessment of their educational system made by the French revolutionaries was more right than wrong. In 1800 Jean Chaptal, the distinguished chemist, was made Minister of the Interior by Napoleon. He set out to reform science education, to effect a closer relationship between theory and practice and to build on the work of the 1790s. His assistant in charge of public education, Roederer, commissioned a study of British mathematical instruction by someone who seems to have known both systems remarkably well. Chaptal and Roederer wanted to reform every branch and level of education, and they were particularly interested in the curriculum in mathematics. Their educational spy reported on every aspect of the British system right down to the use of chalk and blackboards, and he claimed that British education in mathematics was superior.[67] If this was true—and we think it was—the fact is significant and must be seen as one part in the complex story of why Britain industrialized first.

After 1800 the French reformers also established new schools to replace the more democratic *écoles centrales*. These *lycées* were located only in selected towns, favored the children of state bureaucrats, and specifically sought to stimulate industrial development in the region. Vastly expanded, the *lycées* remain to this day one of the more superior forms of secondary schooling to be found in any Western country. In addition Chaptal and Roederer wanted the entire population to receive a primary school education and thus to possess a basic numeracy.[68] Yet significantly the status of theory over practice was retained in the salary structure of the *lycées,* where a professor of physics was paid 2,000 francs a year while the professors of chemistry, French literature, and mechanics received 1,500 francs each.[69] Religion had been dethroned in the curriculum of France's state schools; and science, pure and applied, had been elevated even over the teaching of French literature.

Recently economic historians have come to recognize that the "a small group of at most a few thousand people . . . formed a creative community based on the exchange of knowledge" and they became the "main actors" who ushered in the Industrial Revolution in the West. "Engineers, mechanics, chemists, physicians, and natural philosophers formed circles in which access to knowledge was the primary objective."[70] Out of their complex interactions, inspired as much by the desire to get rich as any other motive, emerged an industrial culture wedded to science and technology as means to an end, power-driven productivity. All the associations and academies we have described facilitated the emergence of this unprecedented productivity. It happened first in Britain, where by the middle of the nineteenth century there were over 1,000 associations for the generation of technical knowledge, with membership of about 200,000. Informal associations, as well as cheap print, made the accessing of knowledge easier and became part of what Joel Mokyr has called the "industrial enlightenment."

By 1800 in Continental Europe, every government had recognized that the British lead in industrial development had to be addressed and that scientific and technical education was required as part of the response. Schools of mining were imported from the German-speaking lands, while the polytechnic university established in 1795 in Paris was imitated throughout Europe. Napoleon's conquests of the Low Countries, parts of Germany, and Italy and Spain brought the French model of technical and scientific education to the whole of Europe. Only Britain and the new American republic retained a decentralized system of education in science. In the early decades of industrial development, the British model seems to have worked reasonably well. When in 1851 Britain hosted a spectacular industrial exposition in London at the Crystal Palace, few would have predicted that within twenty years, its technological and imperial might would be challenged by Continental rivals, particularly by Germany. But those years are beyond the scope of our story. By 1800 experimental science, especially mechanics, and technology, had come to be linked inexorably with industrial development. Being modern, as our Hebrew authors of the mid-nineteenth century knew, required some scientific knowledge. The linkage remains at the heart of modernity now as it is experienced throughout the world.

CHAPTER 3

Popular Audiences and Public Experiments

In the eighteenth century, the growth of what historians are now calling the Enlightenment public sphere can be seen in the numerous sites where the literate gathered. Reading and debating societies, coffee houses, and Masonic lodges proliferated in Western Europe. This turn toward clubbing had much to do with the spread of print culture and the increasing interest of a literate public in a wide variety of issues, some of them highly political and evident in the various crises that punctuated the eighteenth century, notably in France and in Britain.[1] The ever-spreading audience also had an enormous impact on the history of Western science. In the course of the eighteenth century, science burst out of the boundaries set by formal institutions, and the new public made the rise of science both dramatic and self-perpetuating.

In Britain the expansion of science was largely the result of entrepreneurial initiative by lecturers seeking to capture a market, especially among the merchants and financiers who sought any advantage possible in the vast schemes and projects associated with growing industrialism and empire. But everywhere in Western Europe, demand was simply the result of curiosity and religious sentiment.[2] In cities large and small, the market for lectures expanded throughout the century and even, in the case of Britain especially, in the towns and spas of the provinces. British lecturers such as

Jean Desaguliers went to the Continent speaking his first language, French, to continue with the enterprise of spreading the Newtonian gospel.

Lecturers throughout Europe employed demonstration devices. Model machines could explain how the outlandish claims by stock promoters and perpetual motion men were impossible and best avoided. This economic dividend was critical, we believe, to the spread of public natural philosophy throughout the century. The notion that natural philosophers by their demonstrations could explain the differences between machines that might work and those that clearly could not was extremely important in clearing away the fog exuded by unscrupulous, and frequently deluded, promoters. Moreover, the demonstration of machines represented an effort to escape the condescension toward actual mechanists often shown by the privileged and the scholarly. These efforts at application, more than anything else, laid out a career for natural philosophers in the business of mechanical innovation. Such lectures challenged the proposition put by the philosopher Adam Ferguson in 1767 that "manufactures . . . prosper most, where the mind is least consulted."[3] By the end of the eighteenth century, many industrialists refused to believe that arrogance. The experimental lecturers had spread the principles and devices that ultimately made some machines an object of scientific contemplation as well. By 1800 the race was on—ultimately won by Sadi Carnot, a Frenchman—to explain in detail the physics of the steam engine.

Experiment was increasingly regarded as a viable means of attaining credible understanding of natural phenomena and their utility. It also meant the erosion of the power of the solitary and gentlemanly natural philosopher—who would have depended on his social status to assert his claims—and the consequent rise of accessible and available lectures that made the newest discoveries and debates comprehensible to the literate and curious.[4] One of the most powerful transitions that occurred in the history of early modern science was the movement away from the narrow exclusivity of scientific societies promoted by princes or the crown. Experiment opened the door. Experiment demanded replication and witnesses. The scientific revolution thus entered a distinctly new phase characterized by the public disputes of the eighteenth-century Enlightenment.[5] In such a circumstance, it was perhaps inevitable that the apparatus of experiment would be used to empower observers who had sufficient curiosity to take part in the debate over nature's design.

To be sure, there was power in Isaac Newton's dictum in the second edition of his *Principia Mathematica* of 1713 that "God does certainly belong to the business of experimental philosophy." Experimenting could induce piety. But it could do a great deal more besides. Newton did not mean that experiment could also become entertainment, but it did just that for thousands of observers. By midcentury the uneducated or vulgar sort for whom Newton had nothing but distain could be found at experimental demonstrations. He meant to encourage the method of experiment for uncovering elements of God's grand design. Yet, in the hands of ordinary mortals, experiment could be put to a variety of purposes from piety to profit. When the young potter Josiah Wedgwood went into business with a partner, practically the first thing he did was begin an Experiment Book. In its introduction Wedgwood explained that he sought "the improvement of our manufacture of earthen ware . . . the demand for our good increasing daily . . . these considerations induced me to try for some more solid improvement, as well in the *Body,* as the *Glazes,* the *Colors,* & the *Forms* of the articles of our manufacture."[6] As a result of those experiments, Wedgwood china became world famous and the Wedgwood family immensely wealthy.

Experiment turned from elite to public in response to economic circumstances and to popular interest. Experiment was increasingly theater. The more dramatic the experiment was, the more entertaining the demonstrations provided by public lecturers. In such a setting, personal eccentricity could foster success, as it did for the heretical William Whiston. He left Cambridge University because his concept of Christ as a man and not as God finished him with the college authorities. In London, he immediately fell in with Whig champions of broad doctrinal latitude in the Church of England. These political connections translated into employment and opportunity.

Whiston was soon giving lectures on astronomy in the meeting rooms on the bank of the Thames known as the Censorium, which was run by the Member of Parliament and essayist Richard Steele. Whiston was employed at the Censorium to provide dramatic scientific entertainments, probably an offshoot of his lectures at the Whig haunt of Button's Coffee House in Covent Garden or at the Marine Coffee House near the Royal Exchange. There merchants had become familiar with Newtonian science through the lectures of another Whig supporter, and Fellow of the Royal Society, the Reverend John Harris. As coffee houses spread from their apparent origins in Venice to London, they became increasingly fashionable, progressively defined politically by the allegiance of their clientele, more prominent in the dissemination of news, and central to the commercial culture of the eighteenth century.[7] Whiston's entertainments fit the bill. His preaching, on other occasions and settings, that the world was about to end never discredited his popularizations of science. Newton, too, had similar millenarian obsessions.

Scientific Entrepreneurs

Experimental lecturers were entrepreneurs in the scientific marketplace. From the turn of the seventeenth century when mathematical and chemical lecturers had appeared in London, there was always an audience. And when in 1702 James Hodgson left the Astronomer Royal Flamsteed to set up as a mathematics teacher and as an experimental lecturer, the audience subscribed increasingly to the dazzling demonstrations that an array of apparatus could provide. Soon, William Whiston and Jean Theophilus Desaguliers, attracted by the metropolis and its potential audience, were adding to the momentum of public access while the Royal Society still presumed an exclusive authority. Experimental philosophers like Hodgson and the two Francis Hauksbees, uncle and nephew, Whiston, Desaguliers, James Stirling the mathematician, and James Worster at Thomas Watt's Academy were only a few of the entrepreneurial crowd who made a handsome living from the market for Newton's science. They came from diverse backgrounds and they sought upward mobility as well as good livings through the medium of experimental science. The list increased rapidly in the first half of the century. New lecturers appeared, such as the electrical experimenter John Canton of Spitalfields; electrical impresarios like Benjamin Rackstrow; and the inventor Gowin Knight, who sold artificial magnets.

The distinction between the audience for lectures, consumers of instruments, and popular entertainment increasingly blurred. At mid-century, Gowin Knight was able to manufacture magnets; they became commodities in the public scientific culture like the prisms and telescopes that could already be bought off the shelf like toys. The process of making experiment part of a commodity culture was not, of course, uncontested. This was something the Royal Society frequently attempted to police, but to no avail. Knight was able to play on his Fellowship in the Royal Society (FRS) as a promotional platform, as had Desaguliers.[8] The lack of the FRS designation, however, did not stand in the way of William Whiston. His suggested Fellowship in the Society was apparently just too much for Newton to abide as "they durst not choose an Heretick." Whatever one's religious reputation, the FRS label was regarded as a symbol of prestige in the marketplace for science.[9] Thus the lecturer and instrument maker Benjamin Martin by mid-century was actively seeking his FRS. He was unsuccessful, not the least because Fellows opposed to the rage for experiment perceived him as pandering to the masses.[10] Experimental philosophy, at least in its public face, had become too vulgar and commercial for some genteel philosophers to bear.

By the middle of the eighteenth century, in much of Western Europe but especially in Britain, the audience for lectures seemed capable of an unlimited expansion. To some degree, this was precisely the issue. If experiments performed before the vulgar appeared base and inappropriate, they might have been deemed a dangerous encouragement to those who were challenging religious, social, or political authority. Among the important questions to be asked of a growing marketplace in experiments was whether greater numbers of witnesses eroded the difference between the intellectually serious and the merely popular. The popular alarmed philosophers who were dismayed by dramatic demonstrations for pure entertainment's sake. Some demonstrators even appeared to pander to a common and vulgar belief in sorcery—as with the mysterious representations of optical illusions and electrical effects in Paris fairs and boulevards.[11]

Even though entertainment might mean deception, it also drew attention to the serious issues of natural force and mechanical control. The brilliant contrivances, like the singing birds or mechanical chess players of Jacques Vaucanson in Paris, seemed to be stunning achievements. The display in London at the Haymarket Theatre in 1742 of French automata was a coup, and Desaguliers translated Vaucanson's pamphlets for their promotion. The duck that Vaucanson contrived even defecated. These contrivances raised issues of the relationship between animate life and mechanical operations as well as about fraud and fantasy. They also allied science with sentiment and sensibility, with seeing and touching as a way to knowledge. In France such arguments reached their culmination in the 1790s, when the revolutionaries closed the old Royal academies and demanded that education in physics by demonstration and touch inculcate a new morality, one that seized both the body and the heart.

Experimental science, revolutionaries and reformers on both sides of the Channel believed, could bring with it a new clarity. On a practical level it was said to be the only deterrent against fraud and get-rich-quick schemes intended for the gullible. By producing stunning devices, exceedingly highly skilled artisans like Vaucanson also drew attention to the knowledge of mechanical principles that Newtonian demonstrators like Desaguliers insisted would prove the best preservative against frauds. In a manner that would be echoed throughout the rest of the century, he deliberately linked knowledge of Newton's physical laws to practical, economic achievement. In a brilliant tactic, he took notice of the celebrated strongman Thomas Topham, whose feats of strength amazed his audiences in the London marketplaces and in the provinces. In 1740 Desaguliers employed Topham as his demonstrator to show that, with the proper apparatus, even gentlemen might perform with the strength of coal heavers—thus revealing with clarity the natural principles behind statics and lines of force.[12] Vaucanson had simply demonstrated what mechanics could achieve, and Desaguliers added human labor to the imagined improvement that useful science offered. But no good deed goes unpunished. When the great artist Hogarth chose a cleric to satirize in his engraving *The Sleeping Congregation* (with a title that tells all), it was widely believed that he aimed his gibe against the hapless and absent clergyman Desaguliers, who preferred scientific lecturing to the life of the local parish curate.

The promotion of mechanical power was critical to the public appeal of Newtonian science. Even as early as 1704, in his *Lexicon Technicum,* the Newtonian popularizer John Harris described an engine as "any Mechanick Instrument composed of Wheels, Screws, or Pulleys, in order to lift, cast, of sustain any Weight; or to produce any considerable Effect, which cannot so easily be obtained by the bare application of Mens Hands, without such help."[13] Thus, following on three decades of mechanical discussions and displays, Desaguliers in his *Lectures on Experimental Philosophy* (1734, 1744) explored the same territory.[14] Demonstration devices were built—from primitive wedges and elementary pulleys—to explain the hidden powers of market strongmen. So too new electrical machines and models of the Savery and Newcomen engines for raising water were put on display. Mechanical principles were not enough; in an age overwhelmed with invention and investment, it was critical that the limitations of machines be understood if only so frauds could be avoided. Indeed, as Vaucanson later remarked, the reason why so many (meaning, philosophers) relied on theory was that they never had to put any effects into practice. In his view, the "single mechanic has done more for the human race" than all the theoreticians.[15] The reclusive scholar had little chance of winning such a contest. But those mechanics who did understand basic physical laws had much to gain. So too did the investors who relied on them.

One such demonstration device owed a great deal to Desaguliers. He designed it. It first appeared in the public lectures of his nephew, Stephen Demainbray. The so-called Maximum Machine was a model of a machine that showed the maximum power of a man to raise water. The model was not hard to grasp: A toy tavern drawer ran up a stairway (as he might have done to lift barrels of beer from a cellar) and then stood on a platform, which descended under his weight. Attached to the platform by a pulley and rope was a container that, while the tavern drawer descended, raised an amount of water the weight or volume of which could thereby be measured. Thus as gravity pulled the man to earth, the container of water was raised, its precise and measurable quantity known and its distance measured. The critical issue was not the ability of a falling weight to raise water, but rather that there was a *maximum* amount of water

2. The Newcomen engine. One of the earliest steam engines employed by Desaguliers and others; surpassed in the 1770s by Watt's engine.
Courtesy of Thomas Fisher Rare Book Library, University of Toronto

that could be raised over a given amount of time—at least to the point of the man's exhaustion after numerous trips up a gangway. In other words, the idea of a machine limited by nature's laws was grounded in lectures and in demonstration models.[16] In such apparatus lay the essential bridge, devised by some of Newton's disciples, between the intellectualism of the scientific revolution and the comprehension of mechanics and laborers.

Entertainment and the Uses of Nature

The mechanical application of natural laws had never been Newton's objective. By the eighteenth century, the legacy of nature's usefulness to humankind proposed by Francis Bacon resided as much in the rhetoric as in the result. Despite a growing Baconian assertion of the utility of natural law, the usefulness of learning among craftsmen and artisans, and the reprinting of Bacon's works, Newton took a more exclusive approach. He simply did not believe that much could be gained except by the best philosophers. A good example of the tension between exploration and the application of natural law is in the flood of attempts by mechanics and a few cranks to solve problems of navigation that were of crucial interest to the imperial and commercial agendas of the Western European maritime nations. But neither latitude nor longitude was easily obtained at sea with any certainty. From Elizabethan times onward, there were numerous efforts to apply natural knowledge—of the magnetic field of the earth, for example—to the improvement of navigation by magnetic compasses. It is particularly interesting that William Gilbert's Elizabethan assertion of terrestrial magnetism was proposed in a treatise that was explicit in its experimental program—to distinguish it from the criticisms of scholars, those whom he called "lettered clowns."[17] Indeed, more than a century later, a number of the early Newtonians who set out to conquer the problem of finding a ship's longitude ran afoul of philosophers and professors who thought that most inventors of navigational instruments had little to offer.

This was especially true after Whiston and the mathematician Humphry Ditton had succeeded in convincing the British crown to establish the then-stupendous award of £10,000 for anyone who could discover longitude at sea within one degree. The creation of the Board of Longitude in 1714 meant, at the very least, that there was some avenue by which to repulse the many who sought the prize—which proved for many years to be as elusive as longitude had been. Whiston even chased the bounty himself by proposing a fleet of ships to be set near the shore to fire mortars into the air so that the position of dangerous shoals and rocks could then be calculated by the difference in time between the flash and sound of the mortar. Whiston conducted noisy experiments around London to determine the distance from which the scheme would prove effective, but he succeeded only in disturbing his neighbours. The idea was utterly impracticable while at sea, notably in gales or in bad weather when it was most needed, and Whiston became the butt of much satire. Yet, he was among the more credible of the prize hunters. Newton meanwhile was constantly inundated by requests for his approval of harebrained ideas. But his attitude, especially toward those who believed that longitude could be determined by improved time pieces, was that "this improvement must be made at land, not by watchmakers or teachers of Navigation or people who know not how to find the Longitude at land, but by the ablest Astronomers."[18] So, even at a point when the explosion of interest in natural philosophy rapidly gathered pace, Newton wanted to enforce the distinction between a popular, vulgar enthusiasm and what was understood by the most knowledgeable philosophers.

When considering the outpouring of scientific lectures in the first half of the century, we must bear in mind that, regardless of content, they served the purpose of sustaining social stability. It is harder to believe in evil spirits, witches, or prophesies when strange phenomena can be approached experimentally. Newtonian lecturers addressed natural, but still largely inexplicable, phenomena like northern lights or the mysterious and dramatic forces that could be displayed in the laboratory, such as electricity and magnetism. The very fact that Newton's own optical experiments on refraction were highly contested meant that the prism and the spectrum formed part of the early lectures that many an audience must have seen.[19] Not only did Desaguliers lecture on optical phenomena before the

Royal Society but also the subject became a staple of many a Newtonian lecturer's repertoire. Of course, this growing popularity of the topic served to reinforce the Newtonian view that experiment, rather than hypotheses, was the proper way to resolve philosophical disputes. And the fact that increasing numbers of witnesses saw these displays and their demonstration devices made the Newtonian success a public victory. Public credibility was fundamental to the achievement of Enlightenment science. Indeed, disputes between Newtonians and Cartesians on the shape of the earth were to be determined by French expeditions to Lapland and to the New World and by the use of highly accurate British instruments. The instrumental realm of observation had ignited public curiosity.

The Public in Philosophy

In early modern Europe it had become apparent that the uses of science would not be the province of philosophers alone. Yet the philosophers' inability to sequester their knowledge was troubling. From the first steps of the scientific revolution, the alarm of both secular and religious authorities preyed on the minds of natural philosophers from Copernicus to Galileo. As one of Isaac Newton's own disciples, William Whiston, later pointed out when he, like the Boyle lecturer the Reverend Samuel Clarke, came under very close scrutiny for his religious leanings, there was a parallel to be drawn between those who might prosecute a Newtonian in Britain and those who once persecuted Galileo for heresy.[20] The tension between scientific practitioners and those otherwise made nervous by their methods and their conclusions is a constant in the history of science. Thus, it is more than a curious circumstance that books by natural philosophers might be thought to be both alien and irrelevant to the vast majority of readers and, at the same time, profoundly dangerous.

Into this tension Newtonian science stepped in 1696 when plans were put in motion in London for a series of public mathematical lectures to be given free of charge by the Reverend John Harris, an early disciple of Newton. At the time London only offered lectures on chemistry, but the charge for these was substantial and beyond the reach of most artisans. Mathematical lectures, on methods of merchant accounting and navigation, had a potentially wider audience in a commercial capital. The lectures would be funded by the merchant Sir Charles Cox and would take place nearby the Royal Exchange where traders gathered. The link between commerce and public lectures proved crucial to the development of natural philosophy. It laid the foundation not merely for an audience from the Exchange, and from its nearby commercial coffee houses, but also for the affirmation that public knowledge of the uses of the natural world might expand the possibilities of national achievement. The Baconian theme was resurgent.

Newton was not especially sympathetic to such a strategy. He believed, for example, that improvements in navigation, especially in securing an effective means of determining longitude at sea, would not be made by seamen or craftsmen who went to a few lectures but by learned astronomers with the best instruments. To a very great degree, such a belief gave privilege to theory over everyday experience, to philosophers over practitioners. Others in the Royal Society held a different view. Robert Hooke and Samuel Pepys were among those who thought that practical skill ought to be promoted even if sometimes at the expense of book learning. Newton clearly understood this and sought, for example, to strengthen the teaching of practical mathematics at Christ's Hospital Mathematical School, where boys were trained for the sea.

But it was in the coffeehouses of London, in places like Jonathan's or the Virginia Coffee-house, where Hooke and Pepys might be found over a cup, discussing the latest controversy in natural philosophy or mechanical invention, that an alternative was found to the tightly controlled and formal debates of the Royal Society.

Newton, however, had hitched his rising star to the success of the Whigs and hence to the cultivation of gentlemanly and aristocratic patronage in the post-Revolutionary world. He was happier at the podium of the Royal Society.

When Newton became president of the Royal Society in 1703, he acted just like any other seventeenth- or eighteenth-century philosopher in search of patrons. Newton's own career, of course, was one that was very much determined by social and political support. It was because of his political connection to the Whigs that he received his position at the Mint and laid the foundation for his private wealth. Yet Newton also attempted to convince the Crown to help the Society solve the difficulties of its inadequate meeting rooms, which were still in a decaying Gresham College. He drafted a petition to the Queen that proposed a new home for the Society, nearer to the seat of government at Westminster so meetings would be "more convenient for persons of Quality." This location, he argued, would be more conducive to improvements in natural knowledge. Convenience for gentlemen and improvements were not invariably linked, but the argument reveals a philosopher caught in the web of genteel connection that defined early-modern society. Newton had some reason to be sensitive on this point as the Royal Society was increasingly subject to ridicule. Foreign visitors remarked on how the Society appeared to be populated by people of little account—by which they meant of low status, despite the illustrious reputation of its president. If the Royal Society seemed "lower drawer" to foreigners, it was because the learned academies on the Continent were heavily populated with royal pensioners, even with the titled. Given the exclusivity so desired by the Society, venues outside that group blossomed for the promotion of natural and experimental philosophy.

Coffeehouses buzzed with commercial news, of deals being struck, of contracts being made and ships sunk, of auctions conducted and stocks floated. Amid the din and the debate, there were increasing numbers of natural philosophers whose knowledge was marketable, a commodity only slightly more ethereal than any other, full of promise if short on results. These less-exclusive centers of learning, of skill and of trade, also made it possible for the kinds of connections to be made that led to public mathematical lectures and secured an audience. Thus, John Harris came to the Marine Coffee House near the Exchange and gave mathematical lectures initially for free from 1698, but later by a subscription fee to the auditors. This model proved immensely powerful throughout much of the eighteenth century. And it was one of the most important means by which scientific reputations were created in the period.

At the turn of the century, momentous changes transformed Britain. The English monarchy was increasingly secure, defined carefully in an Act of Succession by Parliament in 1701 as a Protestant succession. The Glorious Revolution was beginning to appear irreversible even if the threat from the Catholic Stuart pretender exiled in France was still alarming. Natural philosophy played its part in this stabilizing process. London's burgeoning wealth and vibrancy attracted people of talent and skill who advanced the cause of Newton's science and his experimental method. London was the seat not only of Parliament and the Crown but also of the Royal Society and the Royal Exchange. Between these last two, there was a road well worn by philosophers seeking a hearing and patrons of their own. While the venture of John Harris at the Marine Coffee House began to disseminate mathematics, the ultimate promotion of dramatic experiments turned out to be an even more effective way to attract an audience. Certainly, watching a philosopher reveal the effects of a vacuum on a bird in an air pump was at the same time more riveting and arguably less rigorous than a mathematical demonstration.

Perhaps the best early example of this transition toward the popular was in the career of James Hodgson, a mathematician and assistant to the Astronomer Royal Flamsteed, who left the Royal Observatory at Greenwich for a new career as a mathematical lecturer.

In London, he joined with the instrument maker Francis Hauksbee in public lectures on experimental philosophy that were inspired by the Newtonian philosophy. Not only was experimental philosophy a vast subject but demonstrations required a large array of apparatus of great expense. While the Royal Society appeared moribund and published little of value, Hodgson took the plunge into a public world and soon, despite the expense, could hardly keep up with the demand as a mathematics tutor and as an experimental lecturer. It helped, of course, that his partner in this scheme was Hauksbee, who was not only a demonstrator to the Royal Society but a manufacturer of instruments as well. Hodgson had ambitions far beyond a mere calculator in the employ of the Royal Astronomer.

By 1704, the world of the public lecturers was in flux. Since the time of Thomas Gresham, lectures to the largely untutored had seemed like a good idea. Several attempts in the late seventeenth century had failed because they were dependent on uncertain patronage. In 1704, Sir Charles Cox was apparently no longer willing or able to maintain his commitment to John Harris, who was then forced to rely on subscribers. Hodgson seized the opportunity and announced classes in natural philosophy and astronomy to "lay the best and surest Foundation for all useful knowledge." This had been the dream of philosophers since the reign of Elizabeth. But Hodgson had the advantage of all of those remarkable developments in philosophy and apparatus throughout the seventeenth century, including especially Robert Boyle's experiments with the air pump and Newton's revelations regarding the decomposition of white light into its colors by refraction.

When Hodgson advertised for subscribers to his new lectures, he mentioned two exceedingly significant factors. First, and perhaps most notable, he pointed out that he would demonstrate his experiments using instruments seldom seen outside of the Royal Society. The lectures would proceed at a writing school near St. Paul's Cathedral. Access was then possible to natural knowledge if not to the

Society. Second, the list of apparatus he would apply—an air pump, microscopes, telescopes, barometers, thermometers, a hydrostatical balance, and, undoubtedly, a prism—was exceedingly impressive to his audience. He relied on Francis Hauksbee and on John Rowley, one of the most important instrument makers of the period, because an investment in the required array of apparatus was far beyond the reach of most men. Although entrance to these lectures demanded a substantial fee, this venture meant that experimental philosophy was not limited to those of great private wealth or of considerable social prominence.

Hence, in the reign of Queen Anne and just as Newton settled into his presidency, sites of experimental learning underwent a crucial transition. The access to experiment once limited to the guardians of the Royal Society, for those of "quality and honor," was now simply a function of the willingness to pay a fee. Early modern science entered new spaces. While Hodgson had begun in a writing school, by 1706 he was giving lectures at the Queen's Head Tavern in Fleet Street near the Royal Society, and soon at the Marine Coffee House. Taverns and coffeehouses served the interests of experimental demonstrators. Hodgson's life became a whirl of activity, as he taught mathematics and gave public lectures to pupils from Westminster in the west to the Royal Exchange in the east in the City of London. He was, in effect, overwhelmed by the interest displayed by the public.

The transition from the Royal Society to the wider public fundamentally changed the debate over natural philosophy. The barrier policed by exclusionists had been breached, not by bringing the vulgar into the Society but by bringing philosophy out of it. Hodgson and Hauksbee began to lecture about the very developments then being considered within the rooms of the Royal Society. These involved dramatic experiments on light, especially in a vacuum, the phosphors produced by friction, and the brilliant effects of static electricity and of weights falling in an evacuated cylinder. Many of these phenomena arose out of Newton's own queries, which formed part of his hints for further experiment appended to his *Opticks* of 1704. Hodgson, for his part, provided further demonstrations on the passivity of matter, the density of air, and the effects caused by the evacuation of an air pump. Hodgson's experiments, he suggested, proved that the Vacuists—like Newton and his followers—were correct in their criticism of those who had long believed a vacuum could not exist. He likewise reported on other experiments conducted from the top of the Observatory at Greenwich that suggested that the difference in time between the flash and the report of a large gun fired from the tower might provide a means of determining precise position and distance, a matter obviously of consequence to those who faced this very problem in the navigation of ships at sea.[21] Whiston had been anticipated.

There are many striking examples of how the blossoming audience for natural and experimental philosophy in the early eighteenth century created careers for philosophers. James Hodgson's foray into London's commercial world with his lectures and his apparatus would have been of little lasting significance had he not built on the foundation laid by John Harris. Hodgson, Harris, and Hauksbee soon found that others would join them. Hodgson's lectures at the Marine were taken over by Humphry Ditton, a mathematician and disciple of Newton. But there was another transition that was about to be unleashed, which was as crucial as exploding the boundaries of philosophical societies. This time it meant following Newton's lead from a university college to the increasingly cosmopolitan London.

Unlike Newton, those who came from Cambridge and Oxford could not always rely on patronage to provide them with a living. One of the most remarkable cases in the emergent world of public science was that of the Reverend William Whiston, who had been Newton's chosen successor when Newton left Cambridge. Whiston had recommended himself to Newton in part because of Whiston's speculations over the origin of the earth and intercession of a comet. Such past events, he believed, confirmed the Biblical story of the flood. Whiston had become fairly well versed in Newton's philosophy and was a capable experimentalist. From 1703 at least, the Cambridge mathematician Roger Cotes had been aware of Whiston's philosophical reputation. Cambridge was a hotbed of experimentation. By the spring of 1707, Cotes and Whiston were jointly presenting a course of twenty-four lectures on experiments, in effect a mirror to events gathering momentum in London. Whiston's experience would stand him in good stead, for he would soon challenge the Established Church on its doctrine—probably to the chagrin of Cotes. Cotes might not have known that Whiston was, like the much more secretive Newton, a serious scholar of religious doctrine and was increasingly drawn to versions of the Holy Trinity that varied greatly from those required by the Church of England. By 1709 these views got him into severe difficulties with college authorities in Cambridge and he was expelled from his professorship. Newton must have been appalled by the attention that Whiston attracted.

In 1712, perhaps in search of additional income, Whiston put his philosophical experience to good use when he undertook a series of lectures with the instrument maker Francis Hauksbee. Whiston was busy exploiting the connections he made in London. Very shortly thereafter, prominent Whigs arranged for Whiston to give a series of lectures on mathematics at their haunt of Button's Coffee-house in Covent Garden. Although Whiston did not stay there long, he cemented these connections by his engagement in the M.P. Richard Steele's Censorium, the meeting rooms at York Buildings on the Thames. This provided Whiston with a platform from which to demonstrate his explanations of the numerous astronomical and meteorological events seen in the skies of London in those years.

For someone with convincing knowledge of natural philosophy, the flourishing realm of public lecturing could prove very lucrative indeed. Like his predecessors Harris and Hodgson, Whiston found his way among the merchants of the Marine in 1719. And like some of his well-established rivals, Whiston became especially adept at exploiting events like eclipses, comets, and displays of northern lights; sold printed explanations; and even darkened glasses to view a solar eclipse. His career as a lecturer and publisher on scientific subjects lasted until his death in 1752, by which time his religious passions were increasingly his subject.

The Universe of John Theophilus Desaguliers

The growth of the lecturing empires was rapid in the early eighteenth century, providing many competitors to Hodgson and Whiston. Perhaps the most successful was the Reverend John Theophilus Desaguliers, son of a refugee Huguenot minister, who encountered experimental philosophy at Oxford under the Newtonian John Keill. Desaguliers lectured briefly at Oxford, but by 1713 he too was in London and gave a course of lectures to assist the widow of the elder Hauksbee. Hauksbee's death left the Royal Society without a curator of experiments, a position that would suit Desaguliers well. He soon became a Fellow of the Royal Society and one of Newton's most loyal champions, especially when Newton's optical experiments were disputed by Continental critics.

Desaguliers's career as a public lecturer was a remarkable one. In 1734 he claimed to be engaged in his one hundred and twenty-first course—and he still had ten years to run. By that point, there were many competitors, including Whiston and the junior Francis Hauksbee, the nephew of the instrument maker. Desaguliers was able to exploit his Royal Society affiliation and received numerous offers of patronage from the Duke of Chandos, one of the wealthiest men in England, as well as from the Royal Court where he frequently delivered sermons. Desaguliers recommended himself to the Royal Society because he had developed considerable skill as an experimentalist. He had the support of Newton because he was capable of reproducing the very experiments on light and colors that had been challenged by Continental philosophers. He also had the support of his patrons because increasingly his approach, especially to improvements in mechanics, promoted practical objectives.

Desaguliers's career reveals the way in which the boundaries of the Royal Society became increasingly uncertain in the early eighteenth century. While many lecturers never became Fellows of the Society—notably Whiston, who offended Newton with his outspoken and unfettered attack on Anglican doctrine and who attracted unwanted attention—others found the barrier rather more permeable. The best example of the way in which the public world drew practitioners out of the Society was undoubtedly Desaguliers. From his very earliest days in London, Desaguliers lectured on Newton's famous prism experiments, which were thought to be crucial to his theory of refraction of colors. Of course, these lectures also served the interests of Newton and the Royal Society. In 1715 Desaguliers gave a demonstration of the theory to a French delegation; entertained members of the Dutch embassy, whose secretary, Wilhem s'Gravesande, became his pupil; and revealed Newton's optical discoveries to the ambassadors from Spain, Sicily, Venice, and Russia. Philosophical dispute turned into a great public relations exercise and experimental lecturers like Desaguliers advertised virtually every day in the London press.

Desaguliers claimed the design of hundreds of new experiments to demonstrate mechanical laws. The notion that machines had limits imposed by the laws of nature was a critical issue for philosophers as well as for mechanics. Among his many endeavors, his courses kept him so busy that the Royal Society even complained of the infrequency of his experiments before them. Desaguliers knew very well that the image of the Royal Society, and its apparent unwillingness to support its demonstrators with anything resembling an adequate salary, worked in his favor. He owed much more to wealthy patrons like the Duke of Chandos, whose financial schemes also gave Desaguliers much employment, not to speak of his ministerial role as a clergyman. In his own defense, he told the Council of the Royal Society that he was loathe to provide them with experiments that were not really new, but he could easily give them demonstrations using the array of apparatus that he had already employed in his public lectures. In other words, his priority had increasingly become the paying public. Experimental philosophy was no longer the singular preserve of the gentlemen of the Society. Like Hodgson before him, Desaguliers was a busy man—so busy, in fact, that he neglected the clerical living presented to him by the Duke of Chandos. Complaints were even raised about him leaving a body unburied for want of a proper funeral.

The growing attention Desaguliers paid to practical matters, like the operation of cranes or wagon wheels, was a reflection of the British insistence on the uses of natural knowledge. Of course, this focus had not originated with Desaguliers but had been articulated for a century by the many followers of Francis Bacon. It is interesting, for example, that during the Restoration, when the Royal Society wrestled with its relation to trades and tradesmen, John Wilkins—one of the most noted philosophers of the age and someone who would play a major role in the origins of the Royal Society—wrote his *Mathematical Magick* on the principles of mechanics in a popular fashion. Even then, access was an issue. The powers of machines were to be demonstrated experimentally and made obvious to all.[22]

The aim of some philosophers to appeal to a wide audience was one strain in natural philosophy that surfaced with a vengeance in the early eighteenth century. While the wealth of the state might thereby be increased, it was a significant development that natural philosophy was thoroughly absorbed with the unification of the public good with private interest. In this way, the link between the state and the larger society was reinforced. The generation of wealth by mechanical means was, thereby, a matter on which natural philosophers could provide useful comment. Consequently, many of the lecturers of the early eighteenth century not only relied on Newtonian principles but also deliberately advertised a link between "Mechanical and Experimental Philosophy." Mechanics was the foundation for most of the lecturers who gave Whiston and Desaguliers competition.

Taking stock of the reputation of Newton's philosophy by the middle of the eighteenth century involves more than Newton's great work. After all, if the *Principia* was a difficult hurdle, even for those versed in mathematics, it could hardly have sustained the wide enthusiasm for his philosophy that emerged. But there was a great deal of interest in what Newton had to say and, even more so, in where his promotion of experimental method would lead. Partly as a result, the numbers of instrument makers grew in the eighteenth century, with some of them even selling apparatus "off the shelf" to customers wishing to replicate results. This was an exceedingly important development. It meant that, just as in the audiences for the public lectures, demonstrations of Newton's principles could be increasingly confirmed by numerous witnesses. A Newtonian victory would be a public victory at home to be sure, but also possibly abroad.

Transmission of Newtonian Science to the Continent and America

When the professor of physics lectured to his pupils at Jena in 1795, he made extensive use of the writings of early eighteenth-century Dutch Newtonians, and his note-taking pupils replicated by hand many of the illustrations found in Newton's *Principia*.[23] The surviving notes help us to illustrate two points: the enormous importance of Newton's masterpiece in laying the foundations of modern physics and the role of Dutch philosophers and journalists in bringing

Newtonian science into Continental Europe. Even at century's end, when physics had also come to include new and exciting work on electricity and batteries for capturing it, the *Principia* remained basic as did the use of machines—even if drawn rather than physical—to illustrate the principles of physics. In every language and in sites as diverse as universities and teaching hospitals, or as lively as coffeehouses, teachers began with basic definitions of matter and motion.[24] In French schools after 1750, Newton's system was routinely explicated and the young were told universal gravitation "is a primitive law, and uniquely dependent on the will of the Creator."[25]

The earliest explications of Newton's system occurred in French language journals edited from the Dutch Republic. Ever since 1689, when the Dutch stadholder became King William III of England, the Protestants of the Republic saw the English as their natural allies and protectors against French bellicosity. In 1685 French Protestants had been denied any religious toleration by Louis XIV; and they, like Desaguliers's father, fled in the thousands to the Dutch Republic, England, and the new world. They constituted an international force, and the learned among them disseminated books against French absolutism, translations of English works in political and natural philosophy, and journals in which reviewers told of Newton's achievements or the latest work being done in English science. One coterie of French refugees included booksellers from Germany and a young Dutch natural philosopher, Willem Jacob s'Gravesande. He became the leading Newtonian of the 1720s and 1730s on the Continent.

While in London, Francis Hauksbee lectured on Newtonian mechanics, and in Florence within a few years he could be read in an Italian translation. The influence of Newtonian thinkers in Italy became one conduit that introduced s'Gravesande to Newtonian physics and mechanics.[26] Most important for his career, however, was s'Gravesande's appointment as secretary to the Dutch embassy in London. There he made quick inroads into the Newtonian community, to membership in the Royal Society, and on his return to The Hague, s'Gravesande took up the chair of physics at Leiden University. His *Mathematical Elements of Physics* (Leiden, 1720) began as a Latin work that was quickly translated into English, French, and Dutch. Without doubt, the book was the most important and stimulating physics textbook in the West until well after 1750. In that decade the young James Watt (before his steam engine fame) hired a tutor and worked his way through it, just as decades later students at Jena would be exposed to it. In addition, s'Gravesande worked as a consultant on engineering projects, and one important factor in the rapid spread of Newtonian mechanics lay in the perception and the reliability and applicability of mechanical principles in mining, canals, and power technology in general.

The emphasis on application had roots deep in the practices of artisans throughout the early modern period. It is no accident that the other important Dutch Newtonian, Petrus van Musschenbroek, came out of an instrument-making family that his brother Jan continued for many decades. Petrus joined s'Gravesande as a champion of Newton; and between them by 1730, they made the Dutch schools and universities, Utrecht and Leiden in particular, the best places on the Continent to learn physics and mechanics. Like many British chemists and physicians in the early eighteenth century, the young Jean-Antoine Nollet from France received his education at Leiden in 1736, and he in turn became the leading Newtonian experimenter of the period after 1750. By way of contrast, at the same time the professor of natural philosophy and mathematics at Nuremberg in Protestant Germany used an essentially Aristotelian framework; and, while invoking the experimental tradition advocated by Bacon, Johann Muller treated matter by its forms, earth, air, water, and so on, and sought to explain their movements by reference to those forms. A similar framework could be found in French textbooks on physics of the same period.[27] The process of assimilation for the new science was slow, but it was steady.

Remarkably the effervescence of the Dutch schools and universities before 1750 led to relative stagnation in the next generation. By the middle of the century, the Republic was in economic decline by comparison to its neighbors. Interest in science remained in the larger population and high literacy rates contributed to a process of relative sophistication in matters scientific. As far as we know at this time, the Dutch Republic supported the first scientific society founded by and for women. But innovation in experimental physics and mechanics had passed to France—not, however, without a struggle.

The history of experimentation in the eighteenth century was the history of the promotion of Newton's science. Shortly after his exile in England, the philosopher Voltaire (François-Marie Arouet) proclaimed Newton's success and the decline of Descartes. From London he wrote:

> Everybody talks about them, conceding nothing to the Frenchman and everything to the Englishman. There are people who think that if we are no longer content with the abhorrence of a vacuum, if we know that the air has weight, if we use a telescope, it is all due to Newton.[28]

This was written just as Newton's first generation of disciples were earnestly turning their attention to the latest rage in electrical experiments. Yet, curiously, one branch of the new science—the marvelous phenomena of electricity—makes no appearance in the standard French textbook of the 1730s. Its nature had captured the attention of Newton writing his famous queries at the end of the *Opticks* (the Latin version of 1706), and after 1710 his disciple Desaguliers performed electrical experiments at the Royal Society. If there was one single phenomenon that propelled the new science into the mainstream of popular and learned life, it was experimentation in public on electricity.[29] Very early in the history of electrical experimentation, the assumption appeared that it had relevance to medical practices and possibly curative values. Electricity existed in nature and could also be produced artificially, and the relation between the two types remained a central problem well into the twentieth century. As Benjamin Franklin put it, electricity can "become perhaps the most formidable and irresistible agent in the universe."[30] The earliest original American contributions to Newtonian science came in the field of electricity.

The boundaries of the natural world of the eighteenth century were exploded by exploration as rapidly as by experiment. The expanding empires of the early modern Western European nations (then still including the American colonies) sought to possess nature and to catalog it as well. Like cloudbursts on the shore, ships, mariners, explorers, priests, and philosophers descended on coast after coast, intent on capture and colonization. But empire proved far from unproblematic—even for those whose power guaranteed them certain victory. Empire meant new species, new navigation, new diseases, and new sightings of the rare and even of that only rumoured to exist. For example, when the great French navigator Bougainville obtained sketches in Africa of a quadruped seventeen feet high, the naturalist Buffon told him that this was a giraffe that had not been seen in Europe since the time of the Romans.[31] The vast profusion of nature also meant confusion, full cabinets for collectors, and the cataloging of curiosities, for the empires of land and conquest were also empires of rapidly expanding knowledge. New images and new charts, engravings of the exotic, and scientific instruments—all were tools applied in the possession of nature at the farthest reaches of trade routes and in the European drawing rooms of the curious and the wealthy.

The English-speaking colonists had the enormous advantage of a built-in inheritance that began with Bacon and included Boyle, the Royal Society, and Newton. The useful and the experimental combined with artisanal skills in instrument making meant that, by the mid-eighteenth century, sites of experimental science could be found dotted up and down the east coast of America from the College of William and Mary in Virginia to Harvard College in Massachusetts. In New York Cadwalader Colden became a respectable Newtonian who explicated the master's system. London remained firmly the metropole, and the prize of originality could only be awarded by getting attention in the imperial capital. Emphasis is being laid here on electrical experimentation in colonial America precisely because of its high visibility, its immense international dimension, and the colonial excellence displayed by Franklin and his cohort in Philadelphia.

Perhaps in no branch of the new science did work get translated and circulated as rapidly as with things electrical.[32] Medical doctors, showmen, quacks, and experimental natural philosophers vied for public attention, and claims were made for cures for everything from rheumatism and paralysis to the gout. Nature dramatically accommodated with electrical storms that induced both fear and awe. Climatic comparisons from the Bermudas and the Carolinas delighted correspondents.[33] Electricity could strike people and animals blind, and it could strip paint off molding or melt metals—all effects were recounted to audiences by Franklin and many others in Europe as well. It is not accidental that Franklin's books also contained advertisements for other books on geography, medical problems, and the spectacular (comets and volcanoes). The electrical had come to imply the global, the curative, and the spectacular.[34] With electrical experimentation and shows, science had slipped forever out of the exclusive grasp of the learned and become an acquaintance of crowds, both literate and nonliterate. Its integration into everyday life had begun in earnest. In Italy Volta experimented with a range of substances—silk was among his favorites, so too animal hair, wool, and flax—to test friction and conductivity.[35]

To discover nature by exploration was closely akin to its manipulation by experiment in a multitude of European laboratories.[36] Observation and analysis, mathematics, instruments of navigation and experiment, craft and skill, converged to cope with the explosion of information that followed in the wake of the European empires. The results were more than overwhelming for those accustomed to the certainty that ancient philosophers or modern priests promoted. Between exploration and experiment, the chaos of nature and creativity of classification defined the European Enlightenment.

Witnessing

Experimental credibility gave enormous momentum to the world of the demonstrators and their witnesses. Experiments with the air pump that provided proof of the existence of a vacuum, models of early steam engines, demonstrations by Desaguliers of the capacity of loadstones to make keys magnetic, John Canton's methods of making artificial magnets—all expanded the phenomena to be seen and linked the navigational needs of the state with those of commercial men. Merchants and traders became witnesses on highly contested issues.[37] Many of these experimental displays could be exceptionally dramatic. By the 1730s, among the most notable examples of credibility fashioned out of the spectacular, were the hotly disputed demonstrations of electrical phenomena. Electricity fascinated observers, especially those who saw brilliant static charges in a darkened room. Newton's own speculations in his *Opticks* had proposed a link between electricity, light, and subtle substances that might exist in all bodies of matter. Such suggestions established an experimental program regarding the forces at work between the particles of bodies, or between storm clouds and church steeples.

Simplicity and precision were essential to devices displayed before an audience. Indeed, as we shall see, simplicity in experimental instruments became fundamental to some late eighteenth-century experimental chemists, notably in Britain. By the 1740s, the effort was still being made to design apparatus that revealed basic natural principles, whether of momentum and the collisions of inelastic bodies or the existence of electrical fluids. By such means, in France Nollet was able to attract a broad respectable audience— even involving women, as had Harris and Desaguliers in Britain before him. Nollet, like many others, also doubled as an instrument maker; in fact, his course could be described as a marketing device for instruments he was prepared to sell. Nollet targeted his Parisian market carefully, making his apparatus clear and capable of reasonably simple repair. On the other hand, his apparatus did not always come cheaply. He reportedly sold a telescope to Voltaire and Mme du Chatelet for the handsome sum of 2,000 livres. Voltaire even complained that his investment in Nollet's *cabinet de physique* had ruined him. Of course, Voltaire was exceptional in many regards, not the least of which were being a Newtonian enthusiast in France and his penchant for exaggeration.

The audience for displays of fundamental mechanical principles was obviously significant by the 1740s when explorations of momentum and collisions attracted a great deal of attention. Thus, at the Saint-Germain fair, Sieur Pauliny attended demonstrations of the "forces of attraction, repulsion and suspension." Apparatus was the key. Nollet went further than most and wrote a three-volume work for amateurs so they could manufacture their own devices economically and then easily store and repair them. It is especially remarkable that by mid-century a conscious effort was being made to move laboratory practice beyond the collection of amusing observations. Rigor imposed by properly designed apparatus and experiments meant that discoveries could be made by disciplining the ever-expanding world of sense knowledge. In the dramatic entertainments of magnetic, electrical and even chemical powers, experimenters increasingly subjected themselves to dangerous experiences such as by taking shocks or breathing noxious airs. But as the Abbé Nollet once perceptively remarked, "A course of experiments . . . [is] not a course of experiences."[38]

The essence of the Enlightenment interrogation of nature was transparency and an expanding audience. This is what demonstrations were all about. Like the mathematical lectures from which they arose, experimental lectures were required to provide proof of basic axioms or, in this case, natural principles. And, of course, the mathematical lecturers relied precisely on the merits of such demonstrations to establish their legitimacy. This was not only the antithesis of the credulity, possessed by those who believed in astrology or in prophecy played out in signs in the skies, but it was also an antidote to it. Demonstrations, whether mathematical or experimental, created an atmosphere in which plausibility of any claim could be evaluated.[39] This capacity to assess the validity of any proposition was not only philosophical but also social. The development and promotion of methods of evaluation, notably those of trial and experiment, were critical to the success of the public philosophers. Charlatans had to beware.

Lectures that rendered observations easy also relied on instruments that were readily available and broadly understood. When John Laurence in 1718 spoke to the explanation of the rising and falling of mercury in a barometer and related it to the "Philosophy of Gravitation upon the principles of the GREAT Sir Isaac Newton," he didn't quite have it right. Yet within a few years, the onetime schoolteacher and lecturer Benjamin Martin described the barometer as "the first in Dignity among the modern Philosophical Inventions." By then it was among the group of readily obtained devices like telescopes and orreries that adorned many an Enlightened household.[40] These came to include microscopes, prisms, electrical machines (some even for medical uses), and especially more and more chemical apparatus that turned country homes into private laboratories. The barometer was thus but one example of an increasing array of instruments of decreasing expense and diminishing complexity.

The spread of instrumentation was a crucial development in the scattering of Newtonian principles and in the spread of natural and experimental knowledge throughout the social ranks of eighteenth-century Europe, notably among entrepreneurs, industrialists, and mechanics. Devices built confidence. Desaguliers promoted many of his lectures as a preservative against frauds. In the growing consumer culture of the eighteenth century, this was of utmost importance. But in an emerging industrial culture, of innovation and invention, devices might also prove economically crucial. This was especially true when philosophers played a role in some of the promotions and companies that were based on mechanical contrivances, such as those engaged in drainage or the new steam engines even before the innovations of James Watt. By 1700 steam engines were a fact of life, used largely for draining mines. By 1800 they were powering the new factories, their up-and-down motion translated into rotary motion harnessed to spinning machines. By 1851 six-horsepower engines had become portable, and they were dragged from roadside to farmyard to small factory, ready to do a day's work as needed. They nicely blended into the scenery.

CHAPTER 4

Practicality and the Radicalism of Experiment

The early disciples of Newton captured the market for experimental philosophy. Despite Newton's unease, public science dramatically revealed the principles otherwise hidden behind the imposing edifice to his *Principia Mathematica*. As we know, Newton was not interested in what he considered to be pandering to the vulgar. Newton used higher mathematics to explicate the laws of nature and, according to the mathematician John Machin, "Sir Isaac said he first proved his inventions by Geometry & only made use of experiments to make them intelligible & to convince the vulgar."[1]

Nevertheless, despite Newton's reservations, there were two important aspects of the fashion in public lecturing that gathered momentum throughout the century: first, the emergence of a rapidly expanding public audience for experiments; and second, the demonstration of mechanical contrivances, from simple machines to steam engines, based on Newton's notions of attraction, repulsion, inertia, momentum, action, and reaction. A very large number of these lecturers were intent on showing how machines worked to those who attended out of curiosity and also, especially, to those who might invest in contrivances such as cranes or steam engines for industrial enterprises. In our view, a broad public audience interested in mechanical laws had significant consequences for the process of industrialization in the last half of the eighteenth century.

As early as the 1720s, public science turned increasingly practical and less religious in tone; and by the 1770s, it began to attract social and political reformers.

By the early eighteenth century, it was increasingly clear that the deliberation of useful knowledge was a strategy both Baconian and Newtonian in its origins. Here then was the first apparent knowledge economy. Notably, the lectures of Desaguliers illustrate the turn to the practical, and they had two important objectives. First, they helped to establish his practical credentials and thereby recommend him to investors desperate for advice. His patron, the Duke of Chandos, was involved in so many schemes—from the African Company to the steam engines employed in London by the York Buildings Company—that he required the best advice he could obtain. There were many enterprises that demanded the latest technical expertise and Desaguliers, along with a growing number of his rivals, was in a position to provide it.

Partly for this reason, Desaguliers's second major objective in his lectures was to provide information on the latest achievements in mechanical technology—most notably, on the early steam engines that operated in Britain long before James Watt's designs would help make the first Industrial Revolution. Desaguliers could show the effects of the Savery engine, which emerged at the end of the seventeenth century and which used a partial vacuum caused by the condensation of steam to draw water out of mines. He was soon able to demonstrate the workings of the rival, and much improved, Newcomen engine, which was the first to use a piston driven by steam. Hundreds of these imposing machines were ultimately employed in the British mines. Desaguliers built working models of such devices as well as an apparatus to help clear mines of foul air, a continuous source of danger in mines. He even patented a machine that applied the heat of steam in the boiling or evaporation of volatile substances otherwise prone to explosion if exposed to an open flame. Desaguliers followed these public and practical ventures by lecturing on the expansive power of steam to Fellows of the Royal Society. It is interesting that debates on the nature of the atmosphere also gave impetus to numerous experiments on the capacities of various airs to preserve life—and on chemical tests to distinguish foul from fresh air.

By the 1730s, experiment could no longer be limited to the interests of a few natural philosophers. Once audiences had grown sufficient to support the efforts of many lecturers, then practical matters were as important as philosophical principles. Many copies of individual lectures were published in the first half of the eighteenth century, perhaps so the audience could take away the essence of the presentations. In effect, it was possible to compile a textbook lecture by lecture, as in Desaguliers's *A Course of Experimental Philosophy* (1734, 1744), and each lecture often contained vast mechanical detail. But there were also presentations before smaller, much less well known groups other than the Royal Society, so the members could engage in debate on these same issues. We know of mathematical as well as botanical societies meeting in coffeehouses since the first decade of the eighteenth century.

We can, for example, reconstruct some of the efforts of the so-called Spitalfields Mathematical Society, which was established in 1717 in the industrial east end of London among the Huguenot silk weavers and craftsmen. This was not merely a group of artisans struggling with their chalk over obscure mathematical puzzles. The society's activities broadened throughout the century to the building of an impressive library and the establishment of a sizeable cabinet of experimental apparatus for public lectures.[2] In this case, an organized society appealed to, rather than shunned, the public. This approach, we submit, was exceptionally significant for it clearly meant that Newtonian natural philosophy had found an audience far beyond the great philosopher's imagining. The growth of industrial interests and of the instrument trade meant that experiment and models of machines would become staples of the eighteenth-century lectures. The provision of mechanical knowledge met the needs of would-be engineers and projectors.

The link between experiment and mechanics was crucial to the expansion of public interest in natural philosophy and industrial transformation. Although this has been a difficult association for subsequent historians to make, one of Newton's early biographers hinted that the promotion of experiment could be broadly conceived as an attempt to determine "what nature might do and suffer."[3] Out of this proposition, one might draw the conclusion that practical ends tended to justify experiment beyond the scope of a gentleman's leisurely curiosity. If so, then the early eighteenth century was face to face with the very issue that had once led to the failure of the "History of Trades" in the Royal Society and to the controversial efforts to disparage the interests of artisans and craftsmen. The development of the public world of science forced a resolution of this apparent social conflict without the assistance of the Royal Society. Consequently, when Desaguliers complained in the 1740s about those who were "full of the Notion of the difference between Theory and Practice" he was making a very telling point. Those who saw such a difference were precisely the ones most likely to be misled into believing that inventions might work when they clearly violated even the most basic mechanical laws. This was terribly important in an age that had already been wracked by the stock market crash of 1720, in the so-called South Sea Bubble, when all kinds of imaginary schemes had been floated and fortunes evaporated. Much like our own modern experience, in terms of investment in new technology, ignorance could lead to financial disaster.

Philosophers and Engineers

For experimental philosophers, a major selling point was precisely their knowledge of why machines worked, and why they might not.

Likewise, they stressed that mechanics and engineers needed to possess philosophical knowledge. For example, John Grundy, senior, a land surveyor and teacher of mathematics, proposed that every engineer should "understand Natural Philosophy in order to make his Enquiries just."[4] Shortly thereafter, Desaguliers declared in his published *Course of Experimental Philosophy* that philosophers were actually the only realistic guardians so investors "might not be impos'd upon by Engine-makers, that pretend to (and often fancy they can) by some new invented Engine out-do all others."[5] Thus, the public lecturers began to negotiate the critical space between philosophers and mechanics just as industrialism started to gather pace in Britain.

Where the force of utility began to show its face, the works of Desaguliers and his host of competitors could not easily be ignored. It was left to an early engineer, John Smeaton, in 1747 to challenge those who dismissed what were then sneeringly called "the common Herd of conjuring Philosophers about Town"—as though they were simply making money from magic shows. He proposed that the experimental lectures, of which there were many by then, might prove a worthy employment if natural laws were carefully revealed. Smeaton himself bridged the gap between science and mechanical skill that some historians have asserted. Smeaton, a self-styled civil engineer, was elected a Fellow of the Royal Society on the basis of his reputation as a "maker of Philosophical Instruments."[6] His improvements to the air pump were such that the celebrated chemist Joseph Priestley believed that it would prove important to his experiments on airs. Smeaton was so adept at mechanics that he turned his attention to engine making—but as a designer who could ensure that proper natural powers were not ignored. The actual construction was left to others. Natural principle and mechanical experience were united, especially concerning the effects of such phenomena as friction, momentum, and work, which we know attracted the attention of some of Newton's immediate disciples. In-

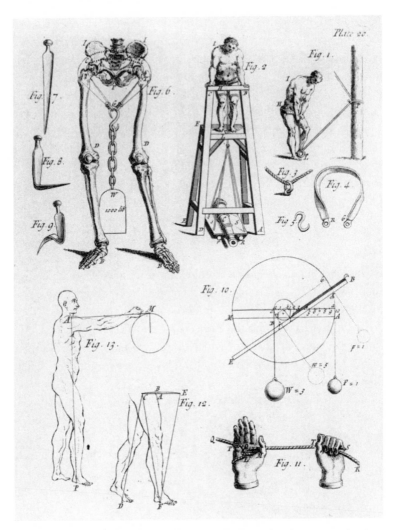

3a & b. The new mechanics. One of the arguments for the new mechanics rested on its ability to augment human strength, with or without new machines.

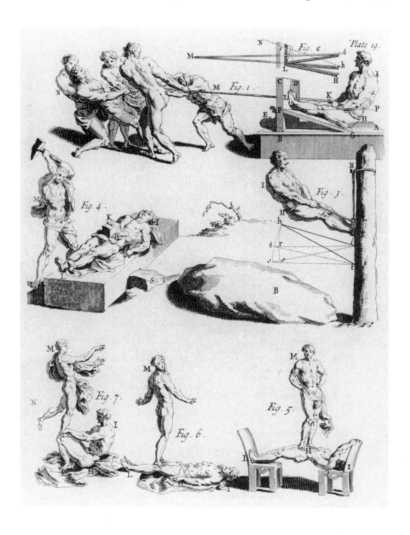

deed, throughout the 1750s, Smeaton conducted a series of noted experiments on wind and water mills and on models made to reflect the operation encountered in full-scale engines. He claimed that the principles that he was able to deduce from such models and experiments would be of great influence as his designs were used over the rest of the century.[7] One can readily understand why, in a century marked by the mathematization of nature, how essential it was to arrive at a calculation by which new mechanical contrivances could be measured and, in that competitive world, compared.[8] It was up to our philosopher engineers to attempt to solve the puzzle of nature's laws when, according to the British historian Paul Langford, "a Nation of Newtons and Lockes became a nation of Boultons and Watts."[9]

While the senior James Watt made a major contribution to the establishment of a universal measure of force by comparing his engines to the power of a number of horses, this actually was a very old notion. As early as 1698, Thomas Savery in *The Miner's Friend* had claimed to be able to build engines equal to the capacity of two horses constantly working together. And the mathematical calculation of power in Newcomen engines was defined in 1721 by Desaguliers's friend, the engineer Henry Beighton, as the result of factors such as the diameter of the piston and the diameter of the pump drawing water, and the amount raised and the depth from which water was drawn.[10] Similarly, half a century later in his appraisal of a steam engine in 1775, John Smeaton calculated its capacity to raise water over a day to a height of 53 feet to be the equal of 400 horses in total. This kind of precise mathematical definition required some sense of what an engine could consistently do, which was not, in fact, easy to measure or even to approximate in those frequently halting machines. The early applications of the Watt engine ran into intense skepticism from investors in mines and factories. Practitioners and proprietors were highly suspicious of the claims of engine makers. The anticipated savings from horses did not uniformly materialize, and the outlay for an engine builder constructing the unmanageable engines often troubled the investors.[11] Nevertheless, horsepower was clearly becoming the standard of significance because it was horsepower that was literally one primary avenue of saving and that would be estimated in a large variety of ways, none of them entirely satisfactory.

We have no hint at how James Watt arrived at his definition, in 1782, of the capacity of a horse to pull 180 pounds 60 yards per minute. But it is clear that during the 1780s, Watt was routinely calculating the capacities of engines at a horsepower standard of 32,400 pounds raised one foot high per minute—though by 1783, he had rounded up to 33,000 pounds. Significantly, Watt was soon describing his engines to prospective buyers in horsepower.[12] It is remarkable that in his own account of the derivation of the standard, James Watt acknowledged a certain amount of fudging and a tradition derived from preexisting mechanical practice: "[Boulton and Watt] . . . felt the necessity of adopting some mode of describing the power, which should be easily understood by the persons who were likely to use them. Horses being the power then generally employed to move the machinery in the great breweries and distilleries of the metropolis, where these engines first came into demand, the power of a mill-horse was considered by them to afford an obvious and concise standard of comparison, and one sufficiently definite for the purpose in view."[13] The notion of horsepower, whatever its real meaning, followed a custom of a lengthy lineage and of the common experience of mechanics in mills and mines.

It may appear that Watt's successful measure was, to some extent, arbitrary. Nonetheless, it did provide the basis for a comparison between engines of diverse types. Boulton and Watt needed such a means of comparison when they sought to replace the engines of competitors or to fight off interlopers who tried to evade their patent. Some engineers early in the first industrial revolution merged experimental method with the application of models. Add to these practices the expanding empire of witnesses in a consumer culture, and deductions might be reached that could be put to actual use. It was the Newtonians, some of whom explicitly advanced the overlap of natural philosophy and mechanics and promoted experiment, who kept Baconianism alive in eighteenth-century Britain.

Experimental Concerns

Let us look at the consequences such courses of experiment might induce. In the early part of the century, the efforts by William Whiston had attracted much criticism from those who feared that lectures before an uneducated audience might spread materialism. His decidedly controversial and unorthodox theological views also did not help matters. Whether the real issue here was Whiston's personality and his aggressive promotion, or the spread of natural and experimental philosophy among paying audiences unconstrained by rank or social position, is not easy to separate. Controversy attracted attention. Increasingly, debates on experiments found their way into the press—so philosophers in dispute might even look to newspapers as an arena in which to settle matters. By the middle of the century, it was not uncommon to find the widely circulated *Gentlemen's Magazine* reporting on issues discussed in the *Philosophical Transactions of the Royal Society*.[14]

Philosophical conflict both engaged the public and could have practical consequences. For example, the efforts to determine an accurate map for French territories required a precise measurement of the length of a degree at various latitudes. The French-Swedish expedition in Lapland to settle the shape of the earth, even with the application of the most exact English instruments, provoked a dispute in the French press regarding whether Newton's notion of a world flattened at the poles had been the right one.[15] Similarly, following a lead set by Voltaire's polemical *Letters on the English*

Nation (1733), thirty years on in the brilliantly written romance of an arduous and celebrated four-year circumnavigation (1766–1769), Louis de Bougainville described himself as merely a sailor and a simpleton.[16] An unassuming popular appeal was sometimes essential in the philosophical marketplace as well as for the readers of travel literature. Philosophy adapted to the stage as well. London's Pantheon was the scene of a celebrated demonstration of the apparent benefit of blunt rather than pointed lightning rods by Benjamin Wilson. He sought advantage in his bitter dispute with Franklin over the best means to protect the King's arsenal at Purfleet. Even the celebrated chemist Lavoisier, in February 1785, carefully staged his theory on the synthesis of water, made from the newly identified gases oxygen and hydrogen, before witnesses at the Paris Arsenal.[17] A scientific public, in the century of the *Encyclopedie,* mattered to philosophers.

Throughout the century, the philosophical world widened rapidly. Thus, we want to emphasize the link between audience and readership on the one hand, and the inducement to exploration of both mechanical and philosophical questions on the other. It is obvious that consequences followed from the elevation of the prestige of mechanical skill. The efforts of the Newtonian lecturers were reflected in the notion put about by Diderot that more "wisdom, intelligence, consequence" could be found in machines for making stockings than schemes for spinning gold.[18] Philosophical fantasy had secured more ruin than riches largely because of the very ignorance the lecturers argued they were able to dispel. If such an ideology were widely accepted, emerging and radical notions of the democratization of knowledge, proposed by the millenarian and republican chemist Joseph Priestley, would have resonated among manufacturers and notably as well would have induced alarm among those whose established social prominence and political power might have been called into question.

Was there then an affinity between the chemical investigations of the experimenters who had once followed the lead of Newtonians like Stephen Hales and those of his successor Joseph Priestley? Did eighteenth-century philosophers help in the accelerating industrial transformation of Western Europe, first exemplified in the English Midlands, and in advancing the parallel demands for social and political reform? Historians have often sensed such connections lurking behind the spectrum of radicalism, republicanism, and reform that transfixed the North Atlantic world in the late eighteenth century. Of course, the industrial revolution pointed not merely to the astonishing growth of new wealth but also to the difficult conditions of daily life that it was also capable of producing. At the same time those with little political power, however wealthy they became from their machines, could point to the necessity of constitutional reform. There was an expanding awareness that social dislocation and difficulties in progressively crowded manufacturing towns urgently needed to be addressed. The airs of the new towns were increasingly foul and that in the factories themselves virtually unbreathable for workers who spent many hours tending the looms and forges. By 1792, a young Scotsman approaching London was appalled that he was "involved in a thick cloud of smoke for the last twenty miles, the wind blowing from the south."[19] Of course, as towns and cities expanded, pollution became far worse and there were few attempts to control it. Inevitably these circumstances raised the social question of what it was that experimentation might do, not merely to accelerate innovation and broaden knowledge, but to address the problems of public health in congested towns and manufactories. For example, Dr. Thomas Percival wrote to James Watt about Watt's rumored invention of "a method of destroying smoke, which issues from fire engines, furnaces, & other works . . . to receive further information concerning a discovery, which promises to be of great importance to the inhabitants of Manchester, who appear to be peculiarly incident to pulmonic affections . . . from the rapid increase of the cotton manufactory."[20]

The study of gases was increasingly important as pulmonary diseases appeared to be endemic and tuberculosis remained a scourge for rich and poor alike. This was at least one reason for the urgency of the chemical analysis of airs in which Priestley and his associates were deeply engaged. In the absence of a germ theory for such diseases, their manifesto as reformers of the medical profession asserted that, by using every possible means, including newly discovered airs like oxygen or nitrous oxide, they might "diminish the sum of human misery."[21] Some believed these gases could be used to treat many diseases, notably consumption (tuberculosis), which was rampant in crowded industrial towns like Birmingham and Manchester. James Watt's own household was severely affected and his daughter Jessy died early of the disease despite his best efforts to secure the most knowledgeable medical advice from the chemist and physician Thomas Beddoes. "Dr. Beddoes's Breath" was one description of the prescription to inhale nitrous oxide and Beddoes and Watt promoted this treatment. Watt even manufactured a portable apparatus for sick rooms. Those engaged in chemical experimentation found that they were frequently called on to address the serious issues of social health, much as was the case in France with Lavoisier. In so doing, the experimental treatments of pneumatic medicine assisted in the blurring of the boundaries between chemists, apothecaries, surgeons, and physicians whose social roles were in significant transition in early modern Europe.[22]

James Watt, in his chemical laboratory, was convinced of the efficacy of the newly discovered airs. He assisted those in Birmingham who were increasingly desperate, often referring them to Beddoes, and helped to organize support among fellow industrialists for the establishment of a Pneumatic Institution by Beddoes in Bristol. He believed that access to medical treatment should have been as easy as breathing air. Pneumatic medicine spread rapidly, especially among those who believed that medical cures should not be limited to those of wealth and standing. Indeed, Beddoes received and published numerous cures that, in the end, turned out to be extremely disappointing; next, he turned his attention to the new promise of medical electricity in the application of the new battery known as the Voltaic pile. By 1792 Beddoes was anxious to absorb Volta's experiments in "a new system of medicine."[23] Perhaps there was a tendency to seize on the latest experimental research rather too readily; but claims of cures were not limited to quacks or to frauds. Experimentation often held much promise, and medical adoption of the latest chemical research was no exception. Beddoes's approach was to disseminate as many reports as he could, thereby attracting even more enthusiastic claims and numerous case studies from medical practitioners throughout the reaches of the British Empire. The Manchester chemist Thomas Henry, while seeking subscribers for Beddoes's Pneumatic Institution, claimed in 1795 that medicine must try all remedies when "the times were so unfavourable, & the prevalence of a destructive Typhus increased the sufferings of the poor to such a degree, as to make every exertion necessary for their relief."[24] And it was at Beddoes's laboratory and clinic in Bristol in 1798 where the young chemist Humphry Davy began to make his reputation on assessing the "prodigious power over the sensibility of the human frame" that some of the new gases and compounds exhibited. Together they tried the effects of nitrous oxide, Davy in particular engaging in a series of experiments on himself so he could adduce the effects. Luckily for him, most of the concentrations seemed to have been low enough not to cause much damage; but Beddoes reported that during a trial of hydrocarbonate gas, Davy made himself very ill "by taking 3 full inspirations of it undiluted, having previously emptied his lungs as well as he could."[25]

The uses of chemistry at the end of the eighteenth century were widely recognized, especially in terms of chlorine, alkali, and ventures in soda manufacturing. Chemical experiments were to a large degree driven by industrial potential and the laboratory was not to be separated from the workshop. [26] The industrialist Thomas Cooper—who, like Priestley, had to flee to America to escape persecution for his political radicalism—declared chemistry to be the most beneficial of all the sciences. Cooper, once a manufacturer in Manchester and member of the reforming Literary and Philosophical Society, stated from America in 1812 that "chemistry is of more immediate and useful application to the everyday concerns of life and it operates upon our hourly comforts [more] than any other branch of knowledge whatever."[27] Thus, by the turn of the century we can perceive the convergence of social reform, republicanism, and the evolving public interest that brought much experimental work to fruition. Whether or not experimental philosophy was intended to further the agendas of eighteenth-century radicals (of which there were indeed many kinds), it remains clear that the utility of such experimental innovation was inherently reformist—and therefore much debated. Hence it was James Keir, the Birmingham chemist, who best expressed a place for the public that we have seen emerge with increasing momentum. In 1789, the very year of the fall of the Bastille and the outbreak of the French Revolution, Keir declared: "To enable these judges, the public of all nations and of all times, to decide with a full knowledge of the question, every view in which the subject can be considered, every argument for and against ought to be presented to them."[28]

Fears of Philosophers

In a world of intense international strain, from industrial wealth and imperial conflict, the concerns of our philosophers came under careful scrutiny. It was one thing for democrats like Priestley or Keir to seize the latest in natural knowledge to further their cause. But it was quite another thing to those who dreaded where this all might lead. Of course, it really did no good for a messianic Unitarian minister like Priestley to light the fuse by exclaiming that the powerful who sat atop the British pyramid had every "reason to tremble even at an air pump or an electrical machine."[29] But a metaphor of philosophy as volatile and incendiary as this was hardly without foundation—not the least in the minds of those who feared Priestley's vision of a world turned upside down.

It is useful to recognize that there had long been a suspicion that philosophy displayed might be philosophy misunderstood. There was always the fear that such a process might encourage those not always well disposed to a power long settled on Church and King. As early as the seventeenth century, there had been much suspicion about where theories of materialism and atomism might lead. In some respects, this is the legacy of the suspicions aroused by religious reformers who had insisted on scripture made accessible, capable of being read in the vernacular languages of Europe without the intercession of priests. In the world of early modern science, the transition from theory and experiment to demonstration and dissemination was the critical evolution. As we know, the Newtonian apostles of the eighteenth century were in the vanguard of this movement. It did not help, of course, that some had reputations for what were then regarded as heterodox and revolutionary religious views.

These reflections against heterodoxy were part and parcel of the High Church backlash against the growing world of public science. This was not news even to Newton's closest contemporaries. His successor at the Royal Mint, John Conduitt, had once cited arguments "against those who prostitute the venerable name of Philosopher to persons who pass their life in making experiments upon the pressure of the air or the virtues of the loadstone, or to the Chymist, or to Free thinkers."[30] To create such associations always had a resonance in the eighteenth century, and they were full of danger for philosophers. During the 1760s, the Reverend William Jones had complained about the fashion for experiments. By then, as we know, Jones was reflecting views already expressed in the press. He

was following the criticisms that had been visited on Benjamin Martin and that herd of demonstrators. One of the most active critics in the 1750s, also imitated and supported by Jones, was the Reverend George Horne, who lamented that the vagueness and "misunderstanding of Sir Isaac's terms seems indeed to have been the seed-plot of all our misfortunes."[31] Moreover, Horne described the courses of lectures then very much in vogue as "a stupid admiration ... [and] a very low and servile employment for a man of genius."[32] But stemming the tide of public interest was a tall order indeed.

By the last quarter of the century, in the age of the American Revolution and fears of social upheaval in Britain, it was increasingly clear to some that a more profound danger was at hand than simply providing diversions for a paying public, however crass that seemed. Of course, the fact that republicans like Ben Franklin had already made a reputation in experimental practice simply reinforced these impressions. To many, the eruption of the French Revolution was the inevitable consequence of a materialist, atomistic philosophy gone mad. In the aftermath of the Terror, the British politician Edmund Burke in 1795 described experimental philosophy as "the principle of evil himself: incorporeal, pure, unmixed, dephlegmated, defecated, evil." The consequence, he announced, was no less than a great unraveling of social order. Look to the shattered remains of France, he demanded, where "you see nothing but the gallows. Nothing is left which engages the affections on the part of the commonwealth. On the principles of this mechanic philosophy, our institutions can never be embodied, if I may use the expression, in persons; so as to create in us love, admiration, or attachment. But that sort of reason which banishes the affections is incapable of filling their place."[33]

The problem for Burke was not France, but what followed. The British establishment was itself under siege and to leave a public philosophy unchallenged seemed to invite the menace of republican sentiment. No social experiments could be permitted, to be induced by the turbulent classes and their spokesmen. In other words, public order was threatened by the public itself. To Joseph Priestley and his democratic friends, this was the aim of progress—an expanding tide of knowledge that, in his *Experiments and observations on different kinds of air,* he argued would ultimately put "an end to all undue and unusurped authority in the business of religion, as well as of science."[34] This vision had to be disputed. Those of more orthodox views would come to essentially the same conclusion—but about the dangers that lurked in a public world of experiment and natural philosophy. It was not just experiments themselves; diffusion made the difference. When the French Revolution was in full force in 1792, in England it was pointed out that "the extension of knowledge beyond certain limits is forbidden by that state of society to which it owes its very existence"—that is, by the "polite and virtuous classes"—or else all authority must be at risk.[35]

Experimental Societies

Dissemination was the key to the republic of knowledge. One vehicle, from the early eighteenth century onward in Britain, was the numerous philosophical societies. They did not set out to rival the Royal Society; but they were an alternative venue for small groups of dilettantes who gathered for discussions on botanical, geological, experimental, or mathematical questions. For some—and certainly for those who did not have the social standing or credentials to gain an FRS—they became centers of sufficiently credible philosophical activity, with their meetings in coffeehouses or taverns, that they provided a substitute for the Society.

One of the most interesting of these societies is the Spitalfields Mathematical Society, which was founded in London in 1717 as a result of the efforts of the mathematical teacher and the naval surveyor John Middleton. Little is known of Middleton or of the early members of the society, which moved from place to place in the

east end of London amid the industries and immigrants but never more than one-half mile from the new Christ Church, Spitalfields. Originally, membership may have been limited to a small group interested in improving their mathematical skills. However, it is very striking that in this area of intense industry, notably silk weaving among the French Huguenot refugees who had fled Catholic France, there soon appeared in the society a few individuals who developed their own reputation in philosophical matters. Perhaps it was their influence that induced the society by November 1746 to purchase "a proper Apparatus for Electrical Experiments" at a time, as we have seen, such matters were much discussed in the press.[36] These interests in the widening fashions of the philosophical world resulted in another description for the Spitalfields group: "Society for Mathematical and Electrical Studies."

Among those Spitalfields fellows who can be identified in this period was John Canton, FRS, who from 1745 operated a mathematical academy, but who made a considerable reputation as an electrical experimenter and supporter of Benjamin Franklin.[37] Similarly, John Dolland, noted for his improvement of the achromatic lens, became a Fellow of the Royal Society and was appointed optician to the King. But Canton and Dolland were exceptional. Most of the members of the society were connected to trades and had little contact with the prominent world of the Royal Society. However, by the beginning of the next century, some of the members had considerable reputations, often as mathematicians or chemists. For example, Robert Porrett was an experienced chemist before he joined the Society and published chemical papers in *Nicholson's Journal*. Similarly, the self-taught William Wilson entered into courses on experimental and natural philosophy in 1798 with Joseph Steevens, an engineer of the East London Waterworks. Wilson designed demonstration devices specifically for lectures, but he also pursued a series of chemical experiments on the properties of chlorine and nitrogen.[38]

The efforts of Wilson reflected the more general objectives of the Spitalfields society concerning the spread of scientific knowledge. In this, they were the exponents of the grand tradition of the public lecturers. The Spitalfields cabinets contained a vast array of apparatus, including what was necessary for electrical experiments but also tools common among lecturers such as air pumps, balances, globes, telescopes, and prisms. By the 1780s it had three electrical machines. Notably this was a society without great resources and certainly without patrons. Yet it managed to maintain a commitment to demonstrations. In 1792, when there was increasing debate on the political consequences of public lectures, the society's rules allowed its members a choice of giving a lecture on mathematics or on "some branch of natural, or experimental philosophy, or shew[ing] some experiment relative thereto." The group took the precaution, however, of prohibiting members from introducing "controverted points of divinity, or politics, into [their] Lectures."[39] Ultimately, they were giving lectures to large audiences for a very small fee, thereby widely propagating the latest in experimental learning. The displays initiated in 1798 were expressly advertised "on terms so easy, as to be within the reach of every individual, who has a taste to cultivate, or a curiosity to gratify." Given the fact that the British authorities were exceedingly nervous about such ventures, this is a singular initiative. And the fee of 6 pence per lecture clearly demonstrated the society's intent to reach as many craftsmen and artisans as might be inclined to attend. Their audiences soon swelled to as many as five hundred.[40]

The Spitalfields lecturers, at the focus of London's manufacturing, also adopted an agenda of utility. In 1804 they claimed that their efforts "tended to facilitate the improvement of the rising generation, and benefited the present by their useful contrivances to abridge the labour of mankind."[41] In part, this claim could be construed as a defense against nervous and suspicious authorities, particularly after the passing of the Seditious Meetings Acts in 1795 and 1799. These acts gave the power to local magistrates to control meetings of over fifty persons and required that the times and places of lectures be provided before licenses could be obtained.[42] Such gatherings were being very closely watched to see whether they were furthering the cause of reform—after all, "reform" could be one meaning of "improvement." The Spitalfields fellow and broker Samuel Gompertz indicated that the Mathematical Society was intimidated by "the information lodged against several members by the gang of informers, who have occasioned so much trouble."[43] Anti-sedition laws, and anti-Jacobin threats, had the consequence of restraining small societies. They had few resources and few connections with which to defend themselves. In any case, by the time of the Napoleonic Wars and a heightened French threat, the official fears had borne fruit. In 1809 the society gave up experimental lectures on the grounds that lecturers were not available—probably simply out of the fear of prosecution, hardly out of lack of interest. It was not until 1817 that the group petitioned the magistrates to let them open their rooms in Spitalfields for delivering lectures for money.[44] By that time there was little chance of a successful revival.

But our Spitalfields mathematicians created only one of many societies that explored the limits of the sciences at the end of the eighteenth century. The most well known is the Lunar Society of Birmingham, which had among its members some of the leading advocates of industry, medicine, and chemistry such as Priestley; Watt; his partner in manufacturing, Matthew Boulton; the industrialist Josiah Wedgwood; and the physician William Withering.[45] To cultivate the seeds sown by lecturers traversing the provinces, like Adam Walker and John Waltire, was possible for schools and societies of all kinds. Thus the Birmingham Sunday Society, to which Priestley belonged, purchased experimental apparatus for use by working men, at least that kind best suited "to a Town of Trade." In 1781 Adam Walker lectured in Birmingham, the heart of steam manufacturing; he was expected to "set people a talking about

Engines."[46] Here is but one example of how the lecturers and the societies were mutually reinforcing in a period of intense interest in technological change.

We need not go even so far as the Midlands to find such interest. In the short-lived society of chemists who met at London's Chapter Coffee House in the early 1780s, there was much discussion of the chemical industry and iron manufacturing. The same interest in science appears at the Chapter Coffee House in the 1780s, the schools offering lectures in Soho like that of Bryan Higgins's from the 1770s, the anatomy school of William Hunter in Great Windmill Street, the Philosophical Institute of Peter Nicholson in Berwick Street, and the Scientific Establishment for Pupils run by William Nicholson. He edited *Nicholson's Journal* and published some of the experiments of the chemists of Spitalfields.

Likewise, in the 1790s, experimental and chemical societies emerged that had something of a medical orientation, such as the Askesian Society, to which Alexander Tilloch, publisher of the *Philosophical Magazine* (and Nicholson's successor) belonged along with Humphry Davy, William Allen of the Plough Court Pharmacy since 1784, and the chemist Luke Howard.[47] And, as we have seen, often a medical interest meant an interest in the social foundations of disease. Even when it could not always be readily accepted, especially given the evaporation of democratic values in the fury of the French Revolution, the specter of reform was virtually impossible to avoid. In a society in the midst of an industrial revolution, where "it is the public which is to reap the benefit of the sluice," this was widely understood.[48]

The promise that induced democratic agitation and the dissemination of knowledge had many proponents in the eighteenth century. Some advocated a program of utility; some went much further to insist on fundamental political reform.[49] After 1789, to others at least, it seemed as though these might come to fruition in the midst of the French upheaval. In 1792, for example, even when it was becoming clear that matters were not unfolding quite as the demo-

crats desired, the chemist and entrepreneur William Howard wrote hopefully to his brother about a "temple of Minerva," that is, about the goddess of Wisdom, arising from the "ruins of the hated Bastille."[50] Such was not to be. One might, nevertheless, have expected some sense of social renovation to be inculcated in plans to give artisans access to the newly created Royal Institution after 1799. It was to the Institution that Humphry Davy repaired after leaving Beddoes in Bristol in 1801. Yet the tone was set, the architect of the new Royal Institution reported, when he "was asked rudely what I meant by instructing the lower classes in science. I was told likewise that it was resolved upon that the plan must be dropped as quietly as possible, it was thought to have a political tendency."[51] Neither industry nor science yet dissolved the boundaries of class.

The circumstances of revolution and war in Western Europe were no laboratory for a crusade of experimental philosophy and the benefits of useful knowledge. Even so, the program of accessibility we have encountered and an emphasis on utility were closely entwined. For example, Samuel Parkes, of the Haggerston Chemical Works in London and a fellow of the Linnean and Geological Societies, published five volumes of *Chemical Essays* in 1815. In them he asserted that the works of chemists like Carl Wilhelm Scheele, Tobern Bergman, and the Cambridge professor Richard Watson had "contributed in no small degree to the information of the public mind, and to that growing taste for chemical pursuits which is one of the characteristics of the present age."[52] While Watson was a Professor of Chemistry, like some of his predecessors at Cambridge, he emphasized the utility of chemistry and took a particular interest in its industrial applications. In his own *Chemical Essays* he asserted:

> The uses of chemistry, not only in the medicinal but in every economical art are too extensive to be enumerated, and too notorious to want illustrating . . . It cannot be questioned that the arts of dy-

ing, painting, brewing, distilling, tanning, of making glass, enamel, porcelane, artificial stone, common-salt, sal-ammoniac, salt-petre, potash, sugar, and a great variety of others, have received improvement from chemical inquiry and are capable of receiving much more. Those who by their situation in life are removed from any design or desire of augmenting their fortunes by making discoveries in the chemical arts will hardly be induced to diminish them by making expensive experimental inquiries, which not only require an uninterrupted attention of mind, but are attended with wearisome bodily labour.[53]

Even in the universities then, supposed bastions of privilege and establishment, there emerged an acknowledgment of the need for the books, lectures, and exhibitions that helped to make the industrial world accessible. It is our contention, as it was Watson's, that such dissemination brought chemistry and industry face to face much earlier than historians have sometimes assumed and in a wide variety of ways.[54]

The expanding universe of chemical and philosophical knowledge eroded the fine social distinctions between philosophers, artisans, and merchants otherwise so readily accepted in the eighteenth century. Perhaps it was the chemist and professor Joseph Black of Edinburgh who faced this issue best. According to one of his students in the 1790s, Black exclaimed:

It is the desire of knowledge, of improvement, of obtaining new facts in any Subject which they take into consideration which distinguishes a Newton . . . or a Boerhaave from the rest of their Contemporaries. They deservedly have acquired the Appellation of Philosophers.

. . . . In short I call every man a Philosopher who invents anything new or improves any business in which he is employed—Even the Farmer who considers the nature of the different soils or makes improvements on the ploughs he uses, I must call a Philosopher, though perhaps you may call him a Rustic one.

Nor am I inclined to give much credit to those men who shut up their Closets in study and retirement have obtained the appellation of Learned Philosophers they in general puzzle more than they illustrate, they are wrapt in a veil of Systems and of Theories and seldom make improvements or discoveries of Use to Mankind.[55]

This is the difference the promotion of experiment had made by the beginning of the nineteenth century. The uses of nature were paramount among many philosophers. New societies in support of a doctrine of improvement—that is, of the cultivation of individual minds as much as of economy—were frequently created. Thus, in London, the City Philosophical Society emerged in 1808 out of lectures by the silversmith John Tatum, but it was soon superseded by the new Mechanics' Institutes. It was left to the nineteenth century to found a series of Mechanics' Institutes throughout Britain. In that national movement craftsmen and artisans attempted to assuage suspicions provoked, by a doctrine of enlightenment and improvement, in the minds of the wealthy. After all, even a new class of manufacturers and entrepreneurs had much to lose if the lower sort were, by rising expectations, induced to revolt. For the industrial entrepreneurs, improvement among the laboring classes could only be seen to be in their own interest. Thus it was absolutely necessary to suppress the tendencies to radicalism and apparent infidelity that had so agitated Edmund Burke and his allies. During the riots that drove Priestley from Birmingham, in July 1791, Matthew Boulton worried that his Soho workmen might be seduced to join the mob pulling down a Black List of houses. Indeed, even James Watt worried about the democratic tendencies of his own son, James junior, friend of Thomas Cooper and a great sympathizer of the Jacobins of the French Revolution. It was not only the landed oligarchy and aristocracy who feared revolt. With industrialists as their founders and patrons, Mechanics' Institutes in the nineteenth century were careful to avoid such implications. Thus, in the rising industrial town of Bradford, for example, by 1832 the Mechanics' Institute took great pains to ensure that it was not seen as "a seminary of disaffection, a school for infidelity, and a nursery for political demagogues and anarchists."[56] It was dissemination of knowledge and skill, and not disaffection, that was being sought. Science in the hands of workers and radicals could bring a new form of power, one that industrialists as well as the state needed to contain.

CHAPTER 5

Putting Science to Work: European Strategies

As noted in Chapter 2, evidence suggests that by 1800 mathematics was taught better in British schools than in French ones. If that was the case, then everyday mathematics, what shopkeepers needed to buy and to sell, or builders and masons to construct, was more widely available. At the secondary school level (grammar schools in Britain), higher mathematics such as geometry and the calculus would also have been taught, though only a minority of boys attended these schools. Even fewer attended the universities for Anglicans or the academies for Dissenters, which were officially excluded from the Established Church. Yet obviously British engineers of all kinds benefited from the advanced training available in the better grammar schools and beyond. The imagined superiority of British mathematical education needs to be noted with the provision that one spy report of 1800 does not prove the case with certainty.

We can be reasonably certain about the different directions that British and French science and mathematics took in the course of the eighteenth, and well into the nineteenth, century. In general—and all such generalizations risk failing to note the exception—after 1750, British mathematics and science turned toward the practical and the concrete; while on the Continent, in France and Germany, emphasis was placed on the development of the calculus as well as on celestial mechanics. More new discoveries in those relatively abstract areas can be credited to Continental mathematicians and scientists than to their British counterparts.

If there was a constellation of innovators in Britain comparable to what there was in Paris from Maupertuis onward, its locus might have been in Edinburgh. There the Newtonian mathematician Colin Maclaurin organized a group with astronomical interests and out of it a society sprang up by the 1750s that included the brilliant chemist Joseph Black. His teaching imparted his own innovations as well as the latest chemical news from France.[1] Yet most of the work of the Edinburgh Philosophical Society was directed at medical problems. The Scottish universities were the real home of Newtonian science throughout much of the century, but on the whole they were centers for teaching and not innovation.[2] A lower level of innovation in the universities does not, however, mean that basic mathematics would necessarily have been less well, or less often, taught in the schools; indeed, from what little evidence we have, the reverse appears to be true.

But scientific innovation, however important, does not concern us in these chapters. Rather we seek to understand the impact of science through its application and its uses in educational and economic settings. The differences in the British and French "styles"—practical versus theoretical—occurred amid increasing competition between Britain and France, and this led to a growing alarm on the French side that the British were winning. As we examine these generalizations about Britain and France—others have also made the same observations—two caveats are in order: First, we know so much more about the British side of mathematics, thanks to the work of the Wallises and others; and second, mathematics in France also had its intensely utilitarian, practical side, but those pursuits just did not overwhelm abstract mathematics.[3]

The turn toward the mechanical and empirical in Britain had many reinforcements. Some were religious critics who cried out that mathematicians were materialists in disguise. Then there were purely economic forces. The market economy in Britain by the 1720s was arguably the most advanced in Europe. It was more unfettered by regulations and state inspections and even by the ability of strong guilds to protect wages. In both France and the Dutch Republic the guilds remained an economic factor while in Britain their power was almost completely broken by the 1660s. By comparison with Britain and the Dutch Republic, France was a maze of regulations, many of them imposed by a government intent on ensuring quality control. This system, devised by Colbert in the seventeenth century, ensured quality at the high end of the market; it was less good at adapting to the need for cheap, fashionable textiles, for example.[4] Yet they were in high demand. In short, increasingly in Britain and also in France, all but the wealthiest landlords operated in or around the market. Simple mathematics, then and now, makes markets work.

The free market meant the use of the land and its produce in capital-intensive ways, with southern England becoming a net exporter of grain well before 1700. Digging and mining to exploit new markets was common in mineral-rich areas, including the famously rich areas of hard coal in the north around Newcastle, where from early in the seventeenth century, entrepreneurs had grown rich and been protected by royally endorsed monopolies. In the west country, mining efforts focused on iron and zinc. One of the great transitions in the history of technology may best be described as the move from wood, wind, and water as the natural sources of power to iron, coal, and steam—in other words, from organic to inorganic sources of energy.[5] For reasons both cultural and economic, England made that transition first, followed rapidly by Belgium and more slowly by France. All these transitions entailed many factors.

In this chapter we dwell on the mental resources and institutions available to the technician-entrepreneurs both in Britain and, most interestingly by way of contrast, in France. To illustrate the British side of the story, we take examples from Manchester, and we look at early industrialists making their mark on the cultural institutions of the town. They sought to create a new generation in their image and likeness. We also take a brief look at the woolen industry in Leeds. Then we survey French efforts, after 1789 led by the state, to transform the educational systems at home and abroad in order to better compete with the British, and through this cognitive change to ensure a steady development of men given to technological and industrial innovation. By the middle of the nineteenth century, state-directed French and German universities, and particularly technical schools, had begun to bridge the gap between science and industry and, if anything, exceeded the British in the arts of application, especially in chemistry.

Before plunging into case studies, we turn to the lives of two practitioners as examples of the British style in mathematics and science. John Rowning learned his Newtonianism at Cambridge, and he then became an Anglican vicar, a scientific lecturer, a headmaster of a grammar school, and the author of one of the most popular scientific textbooks of the century. Members of his family were watchmakers, and he may have practiced the trade before going to university. His *A Compendious System of Natural Philosophy: with Notes Containing the Mathematical Demonstrations* began appearing in sections as early as 1734, and it was still in print into the 1760s. The textbook began with Newtonian mechanics and went—in tried and true fashion—directly to pendula, hydrostatics and pneumatics, optics, astronomy, the nature of light, and at the end to celestial dynamics. Rowning used only Euclidian geometry and, in general, avoided mathematics in favor of straightforward prose descriptions accompanied by detailed engravings. The success of *A Compendious System of Natural Philosophy* meant that it met the needs of its audience in grammar schools and academies. By contrast, one of the first Newtonian textbooks in French relied entirely on mathematical explanations and never mentioned machines or illustrated local motion mechanically.[6]

Perhaps Rowning's accessible and practical book could have inspired the son of a farmer, our second example of British math and science. Thomas Peat became a writing master and accountant as well as a land surveyor and a scientific lecturer in the provinces. He was self-educated (perhaps by books like Rowning's), and he published a textbook in mechanics and *The Gentleman's Diary*, which imitated *The Ladies Diary* (edited by an engineer) and mixed literature with mathematical problems. *The Diary* continued in print under various editors well into the nineteenth century, by which time it had admitted the use of all the mathematical forms from geometry to trigonometry. Remarkably by the mid-eighteenth century, mathematics and mechanical science in Britain had engendered educational entrepreneurs of lesser social backgrounds who made their living exclusively through the practice and dissemination of science.

Religion as a Factor

It is worth exploring the role that religion played in making these differences between Britain and France; then we will address their implications. The development of the market is only part of the story. The other part centers on the relationship between mathematics and religion in eighteenth-century Britain, and that in turn relates to the differences in scientific styles between Descartes and Newton. By the 1660s Newton grew convinced that Descartes's version of the mechanical philosophy would lead to atheism. It offered no place for the work of invisible forces (such as gravity, as Newton conceived it). Increasingly, Newton turned to the Euclidian geometry of the ancients, away from Descartes's analytic algebra, and he got all his disciples to follow the same course. They in turn got all the university chairs in mathematics in the country.[7]

At the same time Newton did not turn his back on algebra, and his mathematical legacy included both algebra and geometry and, of course, the calculus. A follower explained the Newtonian dislike of the Cartesian mathematical method in this way: It substituted "symbols for operation of the mind," and substituted "symbols for the very objects of discussion, for lines, surfaces, solids."[8] Put another way, Cartesian methods were seen to reify things, or turn them into abstractions. Newton's disciples imagined they were developing a mathematics that favored the concrete and the practical. Newton's geometrical proofs worked in the *Principia* for proving universal gravitation, and the ordered Newtonian universe in turn had become a powerful tool in the pious physico-theology that justified the argument for a divine design in nature. Then why did higher mathematics, in particular, come to be seen in some quarters as religiously problematic?

Enter Bishop George Berkeley and his book *The Analyst*, addressed in 1734 to "an infidel mathematician." In it this Anglican bishop applied his own considerable mathematical learning in the service of religion and a philosophy that he almost single-handedly invented, immaterialism. He believed that the material order exists only because, and when, it makes a conceptual impact on the human mind. The effect of this belief was to call into question the naive assumption that there is a simple one-to-one correlation between what sits in the mind and the world outside it.

Berkeley was a fine mathematician who thought, somewhat uniquely, that numbers are merely signs that may or may not have anything to do with the physical world. He argued that while Newton's geometry and his calculus, or method of fluxions as it was then called, were wonderful mental tools, they did not link the mind to the external world. He described mathematicians as mere logicians who might arrive at correct answers while working with false premises. Given the shakiness of their assumptions about the world, Berkeley admonished them that they were "unqualified to decide upon logic, or metaphysics, or ethics, or religion."[9] In short, he challenged the ascendancy of the Newtonians in matters of religion, and he cast doubt on the entire enterprise of basing belief in

Christianity on the regularity and mathematical uniformity proclaimed in the *Principia*. His strategy was to expose weaknesses in
the method of fluxions (we will call it the calculus) and thereby to
claim that, if the calculus relates only to a realm that is ultimately
unknowable, the religious principles of those who promote the calculus will also be shown to be hollow: "If it be shown that Fluxions
are really most incomprehensible mysteries . . . [it] is . . . a proper
way . . . to discredit the pretensions [of those] who insist upon clear
ideas in points of faith . . . [and yet] do without them even in science."[10]

It was as if a wing of the Church—the conservative and Tory
wing—had fired a shot across the Newtonian bow. The new generation of Newtonians, led by James Jurin at Cambridge, the bastion
of the new scientific creed, blasted back and accused Berkeley of be-
having not like a Protestant clergyman but like a Spanish Inquisitor.
Jurin said that he knew Berkeley's true plan; he claimed it was "to
lessen the reputation and authority of Sir Isaac Newton and his followers, by shewing that they are not such great masters of reason."[11] Jurin turned immediately not to a point-by-point defence of
Newtonian science, but instead to the enormous benefits derived
from mathematics, particularly in "the Mechanical arts, in Architecture, civil, naval, and military, upon which the prosperity and se-
curity of this Nation so much depends."[12] But then, as if to effect a
cover for mathematical pursuits, Jurin proclaimed that the calculus
"does by no means supersede the doctrine of Geometry delivered by
Euclid." Almost as an afterthought, Jurin notes that the calculus
hopes to capture, but may never, "the very instant of time that it
[the rectangle AB] is AB."[13] Fundamental to the Newtonian version
of the relationship between Protestantism and science lay the notion that what lies in the mind of the scientist can, with hard work
and talent, be found in nature as it actually is. Knowing the mathematical order found in nature meant a superior knowledge of God's
work.

It would be hard to imagine in the same era such a quarrel between churchmen and mathematicians in Catholic France. There the issue was not what kind of mathematics should be pursued, or whether mathematics conveyed the world as it was and better illustrated the divine hand at work in nature, but rather whether Newton had been right at all. As we saw in Chapter 2, the battle for Newtonian science begun in France by Voltaire and Maupertuis raged into the 1750s (in Italy, into the 1760s).[14] The Catholic clergy who controlled the schools, and in particular the Jesuits, would have to be shown to have backed the wrong science. As we saw, they grudgingly abandoned Aristotle for Descartes, and then dug in their heels.

What made French science so different from British also had a great deal to do with the royal academies and their aristocratic leadership. A Reverend Rowning or a Mr. Peat would be unimaginable in that setting. Instead the French academies encouraged theoretical work among an elite who coveted royal pensions. The battle for supremacy between Cartesianism and Newtonianism occurred within that elite, and only after the victory of the latter did the clerically controlled schools gradually begin to teach the new physics of Newton, and generally only in the 1750s. Deeply theoretical, the curricula of over four hundred French colleges emphasized mathematics and the calculus as well as the theoretical foundations of physics. Little attention was paid to machines or experimental devices.

After the French Revolution and the educational reforms we will examine shortly, the theoretical cast of French science paid off. It was the French-educated Sadi Carnot, a government engineer, who saw the immense importance of the steam engine and in 1824 provided "a complete theory" to explain heat engines of all kinds; he used "the laws of physics . . . to make known beforehand all the effects of heat acting in a determine manner on any body."[15] The British had invented and applied the engines; the French, after their turn to Newtonian physics, explained the principles of physics that made them work. By the 1780s, British industrialists, engineers, and entrepreneurs made science into a way of economic and social life, as we shall see for Manchester. The French state after 1789 made scientific education in every school in the country a formula for industrial development. In the long nineteenth century, and very gradually, the French method that emphasized theory and, after 1789, increasingly also practice paid off. So too did a pattern first developed in the eighteenth century of going to Britain to see the latest inventions and then coming home to imitate and, in some cases, improve on them. In addition, in certain industries such as steam and iron works, British workers and engineers migrated throughout Europe but especially to France, where in 1820 as many as 300 out of 700 workers in iron and steel may have been British.[16] First we turn to Manchester (and Leeds) to see how science was used by local industrial elites to shape the world in their image and likeness. Then we examine the intense efforts of the French state to educate so as to create the industrialist. He was slow to emerge in France, but emerge he did. His British counterpart was always a local invention, rooted in town and countryside, overseen by parliamentary committees but never a creation of the state. On the Continent human agents rooted in their localities applied steam to factories, exploited mines and built railways, but often they needed the powerful backing of state institutions.

Manchester and the Knowledge of Cotton Manufacturers

For a time in the early nineteenth century, Britain led the world in the application of power technology to industry. Indeed the British invented a "new man," the industrialist, a term not in use until the 1860s.[17] Let us look briefly at two such men, early and leading cotton manufacturers James M'Connel and John Kennedy of Manchester. By 1800, when they came on the scene, British educational reformers like the Edgeworths had been arguing that practical, technical, and mechanical knowledge should be taught to children destined for work—at whatever level—in commerce and manufacturing.[18] Cotton masters like James M'Connel and John Kennedy of Manchester were listening. In just five years their firm could expand its energy resources and the output of its steam engines from 16 to 45 horsepower because the partners relied on their own knowledge base as well as that of the engineers with whom they consulted.[19] Just as Boulton and Watt had cut the template for the industrialists who followed them in steam and applied mechanics in general, so too by 1800 M'Connel and Kennedy did something similar in the small world out of which British cotton textiles rose like a phoenix. In cities like Manchester, they gradually imposed their values and their authority in a process that was intensely local but had implications for the whole of Britain. By 1851 and the Great Exhibition of arts and manufactures staged in London, only one British provincial city, Birmingham, led Manchester in the number of exhibitors present.[20]

Family and business papers demonstrate the complexity of knowledge—the involvement in state-of-the-art applied mechanics—possessed by manufacturers like M'Connel and Kennedy who were, unlike Boulton or Watt, not inventors but simply users and reproducers of the new technology. M'Connel and Kennedy took it on themselves to learn as much as they could about mechanics, and the new technology, and to associate with those who had technical expertise. They also got to know every part of the steam engine's construction and oversaw its maintenance.[21] When an engineer from Boulton and Watt's steam engine firm explained how the rotative shaft would connect to the beam of the engine, he was asking them to understand how the vertical motion of the engine would be translated into the circular motion needed to power the jennies. M'Connel and Kennedy had to understand and respond, even correct, the following:

You have here with a plan & elevation of the Engine, nearly according with [the] sketch made by Mr Henry Creighton [from Boulton and Watt in Birmingham] when he called of you. . . . The supports for the P. Blocks of the Main Gudgeon cannot be cast in one piece with the entablature beam, & must therefore be screwed to it. We have drawn the latter *under* the spring beams, but if you wish to have it without the feather on its underside, it may be placed on the upper side, or let into them, and the pillar reach up to it. The cross beams will be much farther asunder than . . . [the] sketch shewed them, but in the present drawings they are as we suppose you intend them: viz the present cross beam to remain where it now is, say 8 feet horizontally from the rotative shaft, and as the cylinder is removed 3.9—the new cross beam will be 7.6 from the old one, from middle to middle. The cross plate might have wings to reach as far as the beams, but query if this be necessary. Or the supports of the P. blocks, might have their base- plate to extend to them, as represented in pencil on the elevation, but we think it looks too much of a thing.
[*Then came the request that makes our point.*] We shall follow your instructions in this respect as also in any other you may point out, as differing from our sketch.[22]

While cotton manufacturers like M'Connel and Kennedy ultimately relied upon engineers like Creighton sent from the steam engine factory to install the machine, they then had to care for it and service it.[23]

As they prospered, industrialists like M'Connel and Kennedy became elite members of Manchester society. Not surprisingly they used their social positions in a township that had grown to about 70,000 souls by 1800 to further promote technological knowledge at large. First their scientific and technical culture enabled their entrepreneurial development, aiding and abetting their wealth. Then that same culture made them genteel. The families of M'Connel and

Kennedy, in particular, became deeply involved in the activities of some of Manchester's most important cultural institutions, especially the Manchester Literary and Philosophical Society and the Manchester Mechanics' Institute, both of which advanced the social image, prestige, and applicability of science and technology. M'Connel was also prominent in the Unitarian chapel life of the city, in particular the energetic chapel at Cross Street. It laid down a cultural foundation in Manchester that wedded science to economic efficiency—and both to liberal, at times radical, politics. By the 1840s Manchester had come to be dominated by cotton manufacturers generally of Unitarian affiliation.

In the next generation the scientific legacy of the city spawned James Prescott Joule, who "with his hard-headed upbringing in commercial, industrial Manchester . . . quite explicitly adopted the language and concerns of the economist and the engineer," and who as a result pioneered the use of electro-magnetism for replacing steam in propelling machinery.[24] At the same time the Tory oligarchy that had dominated the town "was defeated by a coalition of "Liberals," many of whom were still drawn from the network of Unitarian families whose political interests had been first defined in the 1790s."[25] The local power achieved by early industrial capitalists like M'Connel and Kennedy can better be understood if we conceive of them as knowledgeable, and not simply as striving.

But, well we might ask, how did they get to be so knowledgeable, how did this particular kind of scientific and practical knowledge come to be theirs? Born and raised in rural Scotland, M'Connel and Kennedy migrated to Lancashire and became, within fifteen years of the start of their partnership in Manchester, members of the region's elite sociocultural institutions.[26] Common assumptions would seem to dictate that the technical knowledge of these early industrialists was primarily of an untutored, artisanal sort with invention by tinkering superseding abstract knowledge of scientific or technological principles.[27] In this view "practical knowledge" is largely divorced from "theoretical" or "abstract" knowledge. But technical knowledge during the late eighteenth and nineteenth centuries consisted of multiple skills of varying degrees of abstractness.[28] Following the efforts of Newtonian lecturers, a new "technical literacy" came into being along with new manufacturing technologies and included, in addition to traditional alphabetical literacy, the ability to make mathematical calculations of increasing sophistication and the ability to read and understand technical drawings and models.[29] Somehow M'Connel and Kennedy acquired such sophistication. We cannot document exactly where or when the acquisition occurred, but we can demonstrate its presence. According to M'Connel family memory, James hiked the four or five miles from his home to New Galloway to study at the parish, while Kennedy got lucky when he encountered, just before leaving Scotland for England, a young teacher who "opened my mind to the beauty of mechanical pursuits, and gave me some ideas of connecting a few causes together to produce a desired result."[30] Significantly, Kennedy attended the natural philosophy lectures of John Banks, who was then offering a course in Preston. Banks gave lectures throughout the Midlands and at Manchester's College of Arts and Science.[31] Neither M'Connel nor Kennedy ever pursued natural philosophy as an avocation—as manufacturers like Boulton and especially Watt did—yet they proved to be well disposed to take advantage of the burgeoning interest in technological innovation. The breadth of interests Kennedy displayed in the essays he wrote for the Manchester Literary and Philosophical Society demonstrate his technical learning, and we know that emphasis on technical skill was reiterated again and again in the Kennedy household. "Hearing this advice so constantly repeated by my mother, that we must learn to work with our hands, and seeing also how difficult it would be with our slender means to get on in the world, I at last screwed up my courage to say, I would leave home and become an apprentice to some handicraft business."[32] Kennedy took the message of application and virtue to heart and throughout his life remained, like Banks, concerned not only about the state of his own knowledge but also that of the laboring classes.[33]

As their business prospered, M'Connel and Kennedy soon found themselves among Manchester's wealthiest citizens. With this wealth came social responsibility. During the first decade of the nineteenth century, both men joined the Manchester Literary and Philosophical Society.[34] In 1803 they became inspectors of the Manchester Infirmary.[35] During the 1820s they donated money to, and sat on the board of, the Manchester Mechanics' Institute, an organization dedicated to spreading precisely the kind of knowledge employed by the partners.[36] Furthermore Kennedy served as a commissioner on the Provisional Committee in charge of the Manchester/Liverpool railway project and as a judge in the Rainhill locomotive engine trials of 1829. M'Connel served as a commissioner on the building of roads.[37] Given the importance and authority of the positions they held, we can see technical knowledge played two roles. It provided a means for increasing wealth and it acted as a form of cultural capital that industrialists held in common with engineers and artisans, physicians, and practicing natural philosophers.

Contacts that M'Connel and Kennedy made within Manchester religious life could also translate into other social realms—for example, into entree to the Manchester Literary and Philosophical Society. That group first met in a back room of the Cross Street Chapel and included numerous prominent members of the congregation among its members.[38] Founded in 1781 by twenty-four members of Manchester's professional and religious elites, the Lit and Phil focused on art, literature, and natural philosophy. At its founding, the Society's membership included physicians, ministers (Anglican and Dissenting alike), lawyers, and merchants and manufacturers, many of whom took an active part in the literary life of the society, publishing papers and giving public lectures.[39]

The published memoirs of the society include papers on natural, experimental, and moral philosophy, as well as literature, history, art, and education. Although the "useful arts" were considered pre-eminent among the society's interests from the start, it was several years before topics related to manufacturing and technology stood with equal stature alongside topics related to the liberal arts. While the early published memoirs primarily represented the interests of local religious figures and physicians, later volumes included the work of engineers, merchants, and industrialist/manufacturers like Peter Ewart, a member of the Mosely Street Unitarian chapel, who presented on the mathematics of motion.[40] He also assisted woolen manufacturers in Leeds in installing their steam engines. Out of the society sprang a College of Arts and Sciences. It lasted only a year or two but became something of a precursor for the Dissenting academy that opened in 1786, supported and operated by local Unitarians. Only after protest from non-Unitarian members of the nonsectarian Lit and Phil was the academy declared wholly independent of the society.[41]

Of the two partners, John Kennedy made the most out of his connections to the Literary and Philosophical Society. Between 1815 and 1830, Kennedy read at least four papers to the society, all of which were published in the society's proceedings. In them Kennedy addressed themes that had become central to industrial life: The development of the cotton industry, the effects that machinery was having on the laboring classes, the social consequences of the poor laws, innovations in cotton machinery, and the economic implications of the exportation of British-made machinery to the Continent.[42] For the most part, Kennedy's intellectual development closely followed positions staked out by Manchester's liberal industrialists. For him, innovations in cotton machinery drove the growth of the industry and represented the diligent and creative work of its practitioners. Diligence and application also reflected the moral health of the new industrial class. Indeed, he argued that

the rapid growth of industry in Lancashire was "chiefly to be as-
cribed to the great ingenuity and the persevering, skillful, laborious
disposition of the people." "In these qualities," he continued, "I believe they surpass the inhabitants of every other part of this island,
or of the whole world."[43] Thus industrial progress was itself a sign
of social virtue, and the appalling conditions of the working classes
a matter solely of local concern and ultimately to be corrected by
market forces.

With such arguments in hand, Kennedy, now a Liberal ideologue,
defended the new factory system from its critics: "The frequent
complaints, both in public and private, against the manufacturing
system, certainly demand an impartial investigation," Kennedy declared in a paper read before the Lit and Phil in 1815, "and none
are more called upon to take a part in such discussions than those
who are interested in manufacturers."[44] Kennedy argued that, far
from contributing to the deterioration of morals, the creation of
large factories and regular work hours had "good effects on the
habits of the people. Being obliged to be more regular in their at-
tendance at their work, they became more orderly in their con-
duct, spent less time at the ale-house, and lived better at home. For
some years they have been gradually improving in their domestic
comforts and conveniences."[45] This public defense of the industry
paralleled concerns that M'Connel and Kennedy expressed in their
business correspondence.[46] Attempts to investigate and regulate the
conditions of workers in factories, for example, came under strong
opposition from manufacturers like M'Connel and Kennedy, who
considered Peel's efforts "a very dangerous interference" and liable
to have "consequences [that] may be very injurious to all large
Manufacturers of every description."[47] The Unitarian emphasis on
personal freedom could cut in decidedly self-serving ways.

For cotton masters like M'Connel and Kennedy, membership in
the Manchester Literary and Philosophical Society translated into
politeness, gentility, and charity—all impulses that led to the found-

ing and operation of the Manchester Mechanics' Institute (founded 1825).[48] In addition to making substantial donations of around £600, both partners sat on the Institute's board of directors, which made decisions regarding Institute buildings, notably the purchase of experimental apparatus, the setting of the curriculum, and the hiring of tutors and lecturers.[49] In a highly ambitious project, the Manchester Mechanics' Institute ran an elementary school for reading, writing, and arithmetic, and it offered workers numerous lectures and lecture courses devoted to topics in mechanics, chemistry, natural history, mechanical and architectural drawing, and geography. By appealing to the ideal of self-improvement of both the working and middle classes, the Institute provided an important means for popularizing science and encouraging specialized knowledge among factory employees. Even when it was not the focus of a course as a whole, practical knowledge stood out as a major component of the curriculum. As stated in its published rules, the Institute's established goal was to enable "Mechanics and Artisans, of whatever trade they may be, to become acquainted with such branches of science as are of practical application in the exercise of that trade; that they may possess a more thorough knowledge of their business, acquire a greater degree of skill in the practice of it, and be qualified to make improvements and even new inventions in the Arts which they respectively profess."[50]

In the first lecture course on mechanics offered by the Institute, the lecturer covered a host of topics central to industry. Two of the twelve lectures covered the design and operation of gears (particularly with respect to their use in mills), the arrangement of their teeth, their operation in couplings and governors, and their utility for equalizing motion. Of the four lectures devoted to the concept and application of force, three included discussion of wind and water mills, and steam engines.[51] In subsequent years the Institute offered courses and lectures ranging from Mr. Adcock's lectures on the "Elements of Mechanism, as applied more especially to the

Construction of Steam Engines," Mr. Hewitt's eleven lectures on the "Geography of British India, China, Central Asia, Turkey, Egypt, Arabia, Isles of the Indian Ocean and Northern Regions," Mr. Sweetlove's "Philosophy of the atmosphere," Mr. Bally's course on "Plaster & Wax casting, modeling, etc.," to Mr. White's "Power-loom weaving."[52]

By consensus among both contemporaries of the 1840s and historians, British basic and primary education had failed on a national level to make a more educated industrial and laboring class.[53] By that decade there were over 600 mechanics' institutes, but most lacked a good primary school foundation in their localities. In localities like Manchester, possibly a far more optimistic picture can be presented. For M'Connel and Kennedy, and other prominent Manchester cotton and steam industrialists like William Fairbairn and Peter Ewart, who also sat on the board of directors, the educational mission of the Institute embodied the importance they themselves had come to ascribe to mechanical knowledge and formal learning. In effect, the Institute also elevated the status of practical knowledge to that of natural science in general, both by including topics of practical importance in courses on natural science and by offering independent courses devoted to practical skills like mechanical drawing and machine operation. The way M'Connel and Kennedy had outfitted their own minds had become a hallowed truth: Technical and scientific knowledge were mutually supporting and intertwined. Although the degree of success varied widely from institute to institute around the country, the Manchester Mechanics' Institute for laboring men proved to be a resounding success for several decades, in large part because it attracted the participation of workers and manufacturers alike. To the working class, it offered an opportunity to improve basic skills through the knowledge of "scientific principles"; and to the industrialist, it offered another opportunity to shine as cultural and civic leaders. Neither opportunity was intended to change the social or economic place of workers. Yet technical learning offered social mobility, as Kennedy and M'Connel would have been the first to tell the workers.

Kennedy and M'Connel should not be seen as eccentric or unique. In Leeds as early as the 1770s, the new woolen industry was taking shape in a conversation between manufacturers and engineers around bringing coal to the town via a new canal.[54] Then in a leadership role that prefigures that of Kennedy and M'Connel, Benjamin Gott emerged as the entrepreneur in the town who possessed significant technical knowledge. In 1792 his woolen and worsted manufacturing firm consulted with Boulton and Watt about installing a 40-horsepower steam engine, and Gott, its most mechanically proficient partner, became a consultant in the region on engineering problems. He also pioneered the use of steam in the process of wool dyeing.[55] He became an expert on a hydro-mechanical press, or Bramah's hydraulic press as it became known, a large and complex piece of equipment introduced late in the century, which required an understanding of levers, weights, and pulleys, as well as air and water pressure and which was used to imprint patterns on textiles.[56] He carefully compared the relative merits of prototype machines offered by rival manufacturers of the device, but the machine met fierce opposition from his workers and may never have been systematically used for years.[57] The hydro-mechanical press raised an enormous weight to a small height by using a strong metallic cylinder, accurately bored and watertight, which was connected to a small forcing pump. By means of valves, pumps and levers, cisterns, and water pressure, 400 pounds of pressure was accumulated and then released.[58] The press was to be used to apply patterns to worsted just as it had been used in applications to cotton. It called on just about every principle learned in Newtonian mechanics, as taught from Desaguliers to Dalton, and no semi-literate tinkerer in the country could have made sense of it. The knowledge economy advanced in the textbooks lay embedded in the cotton and wool factories of the 1790s.

The Gott firm and family also became leaders in the civic and industrial life of Leeds. Just like the Boultons and the Watts, the M'Connels and the Kennedys, the Gotts and their local equivalents, the Luptons, Marshalls, Adams, and Walkers, established themselves as leaders (or proprietary members as they were called) of a Philosophical and Literary Society. In 1821 the opening language it used valorized science and the industrial order: "The thirst for improvement gives an exaltation of character . . . produce[s] the works of genius and the discoveries of science."

The lives of M'Connel, Kennedy, Gott, Bouton, and Watt also highlight the cultural and intellectual dimension in the story of early British industrialization. In essence, they demonstrate that both scientific and technical knowledge informed business and social relationships. At first the knowledge helped provide a common language that made it possible for manufacturers and engine builders to communicate, to build and use machines of increasing sophistication and complexity. Then the knowledge provided the veneer of gentility necessary so manufacturers could meet with established professionals—especially medical men and religious leaders—who had risen to a significant place in elite society. The history of early British industrial development attains greater nuance and sophistication when technical knowledge, and its cultural matrix, are restored to the story. The ascendancy of an industrial bourgeoisie can be seen to be a process that involved hard capital to be sure, but also scientific knowledge of an applied sort, a distinctive form of cultural capital.

French Educational Reform after 1800 and the Making of Industrialists

Within scientific circles on both sides of the Channel, by 1800 the dream may have been one of cosmopolitan cooperation, a sharing of chemical and mechanical knowledge.[59] But the harsh political re-

tury.[86] In a cotton town like Rouen, industrialists were meant to be highly literate as well as scientifically knowledgeable. They might have come to be seen as more genteel than their Manchester counterparts, but it is by no means clear that they would have exercised the kind of authority over local institutions that entrepreneurs like M'Connel and Kennedy had achieved.

The late 1820s and beyond saw further complaints about the educational system—"those destined for industry have not found . . . the elements of instruction that they need." The revolutionary ferment of 1830 led to the creation of a central school for arts and manufacturing in Paris and one for commercial life in Bordeaux. Indeed other towns competed to have their sites chosen for these new schools.[87] The ferment of the late 1820s even meant that the state school for engineers had once again to be reformed. After the 1790s its pupils had at least begun to think about the needs of entrepreneurs, but conservative forces at work in the period from 1815 to 1830 had been more interested in achieving order than in fostering industry. In 1830 and again in 1839, the whole curriculum was revamped. The reform arose out of yet another revolution in France's troubled political landscape, and once again progressive reformers came to power who favored industrial development.[88] After the reform, the engineers were to be taught political economy and to understand the actual, as opposed to the theoretical, construction of buildings and machines. They were expected to see themselves as "men of action rather than of words."[89] In the 1840s the school for state engineers had its premises expanded to include a large gallery for machine models. New textbooks were adopted that hammered home the point about application and science.[90]

Arguably all the French efforts of the first half of the nineteenth century were too little, too late. As we are about to see when we turn to the exhibition of 1851, the British industrial empire remained formidable until well into the second half of the nineteenth century. The institutionalization of science within the British universities also gradually corrected the tendency so visible there during the eighteenth century to separate practice from innovation. Theoretical and "pure" science once again achieved curricular preeminence. Yet another, localizing reality also took shape in nineteenth-century Britain. Industry and application, the sciences of utility and innovation for profit, became the projects of certain places and not others. Throughout the nineteenth century, for example, London and Liverpool remained largely commercial; Manchester and Birmingham were the apparent centers of industry, of new men like the Watts and the Kennedys.

Industrialists became fabulously rich, but they were never seen as entirely genteel either by the old commercial classes or by the highborn and landed. By contrast, if ever so slowly, France built a national consensus about the value of industrial life and the necessity for scientific literacy. Gradually, and only after considerable political turmoil, the French state come to stand squarely behind a commercial and industrial bourgeoisie, one that it had worked so hard to shape and mold. The issue of British industrial decline—a phenomenon with a long historical shadow that surfaced dramatically from the 1960s until the 1990s—is far beyond the scope of our expertise. Yet it must be remembered that all decline, so-called, is relative to another country's share in global wealth. Part of the explanation for Britain's global decline between 1880 and 1980 may lie in some of the educational rigidities we have described in this chapter.

The failure to create a national consensus and culture based on the broad virtues of utility—as opposed to local strongholds of industrial culture based on those values—may go part of the way toward addressing the relatively steady industrial development that has characterized the national histories of the other European powers. The very inventiveness that owed so much to English science from Bacon to the Newtonians got institutionalized by the Continental educational systems controlled by the national state. The process took many generations and had many fits and starts. But in the end, by 1914, the French tortoise caught up with, and after 1960 surpassed, the British hare. To this day, a network of eight national schools produces annually about a thousand engineers who lead French industry. They are the descendants of an educational system slowly put in place after 1789.

By the 1870s British reformers were demanding educational reform on the French model. As early as 1867 the Prince of Wales, no less, said "that superior industrial education is making French, et.al. surpass the English in inventiveness." Within the decade they had to face the challenge of a unified and technically proficient Germany as well. Gradually British educators and industrialists began to embrace the necessity for technical education particularly at the secondary level, and reform became what one historian has called an ongoing and "long battle."[91] Arguably educational vitality remains an issue in contemporary Britain, which has the highest rate of failure to complete secondary education of any EU country. One reason that the school-leaving rate remains so critical is the inability of young men and women to achieve a high level of numeracy and literacy when they leave school at 16. Modern technology favors the mathematically nimble.

The Exhibition of 1851

In 1851 none of the problems associated with decline were evident when the Prince of Wales, Albert, husband of Queen Victoria, opened the Great Exhibition at the Crystal Palace, a vast glass hall built especially for the event. The largest enclosed space on earth, it held over 100,000 exhibitions of industrial goods, techniques, and arts and crafts from all over the world.[92] The French had pioneered the industrial exhibition—the forerunner of the modern World's Fair—and opened the display of French products for the first time in 1798. On that occasion the Minister of the Interior invoked the spirit of Francis Bacon and reminded the audience of his call to gather and examine all the things found in nature. At the state-sponsored exhibition, hand-manufactured and high-quality French products from the entire country were on display. Very little mechanized industry appeared in Paris in 1798; cotton-spinning machines but no steam engines were to be seen. Yet the success of the event led to its repetition—in 1801, 1806, and, after 1819, regularly and systematically. In 1844 in Paris the exhibition was international in scope for the first time, and in 1849 the city of Birmingham attempted to imitate the French model. The British efforts, including

4. The historical dimension of the Crystal Palace Exhibition. The exhibition aimed to show the development of industrial machinery over time, here illustrated by the cotton power loom.

Courtesy of Young Research Library, UCLA, *Official catalogue of the Great Exhibition of the works of industry of all nations*, London, Spicer Brothers, 1851

an exhibition in Leeds in 1839 that drew 200,000 people, had been entirely local in inspiration and private in financing.

The exhibition of 1851, by contrast, had royal endorsement, a nationally known committee heavily weighted with Whig dignitaries, and it was strikingly mechanized. To a man, the committee subscribed to an ideology that was "progressive" and that wanted to make engineering, science, and technology endlessly innovative.[93] Indeed the event itself caused increased attention to be paid to the entire issue of technical education and industrial competition with the Continent. Here was a site where manufactured wares from all over the world—including the rapidly industrializing America—could be seen and compared. The purpose of this exhibition of unprecedented size was to show the world the depth and breadth of British industrial development. It also aimed to "exhibit the beautiful results which have been derived from the study of science."[94] Half of the space was devoted to British devices and design; everyone else fitted into the remaining space according to a complex formula and the level of demand from abroad. It was a smashing success.

With funding solely from private initiatives, the British exposition was planned and executed by contributions from around the country. Every town and city formed committees and sent donations; and at the national level a Royal Commission appointed teams of judges to evaluate the proposed exhibitions sent by manufacturers. The dream of the exhibition entailed nothing less than "the realization of the unity of mankind."[95] The exhibition sought to display the achievements of science that "discover[ed] the laws of power, motion, and transformation," and to showcase how "industry applie[d] them to the raw matter which the earth yields us in abundance."[96] Raw materials such as coal and cotton from the field were to be seen as were countless steam engines—for ships, railroads, and factories, small and large—as well as spinning machines, tools of every trade, and fine hand-worked leather and dyed cloth.

By far the most important series of displays centered on "machines for direct use," that is, machines that harnessed the laws of power and motion into the production or movement of goods. It is possible to gauge the relative size of each city's industrial base from the number of exhibitors it sent to London: Belfast had 53 while Leeds sent 134, Manchester 191, and Birmingham a huge 230. With its vast population, London led the count but, relatively speaking, its manufacturing sites were less mechanized than what could be seen to the north.

In 1851 fine handwork still counted enormously in the world of manufactured goods. Indeed the taste displayed in furniture and material culture in general was overwhelmingly aristocratic. The machines with their sleek metal, often displayed chronologically by the stages of their developing complexity, competed for attention with fine and ornate china and vases that only the very wealthiest could afford. Most of the published catalogs, aside from the official one, devoted overwhelming attention and space to the *objets* that would grace the homes of the high born.[97] Visually the Crystal Palace displayed British social reality: Mechanized industry and science were the tools by which new national wealth evolved. But in matters of style, taste, and personal consumption, birth and social place still defined the ideals. At the opening ceremony, the prowess of the British Empire was also put on display, but few of the colonies had anything other than minimal space allotted to them. Even so, American entries gained considerable attention.

The compilation of the exhibition's official catalog tells us much about the interface between science and industry in Victorian England. Industrialists sent their machines with descriptions, but the working of the devices had to be replicated on the floor of the exhibition and also clearly explained in the three-volume massive catalog that accompanied the show. "The occasion called for a large amount of peculiar knowledge—knowledge not to be gained by study, but taught by industrial experience, in addition to that

5. Portable steam engine. Industrial development reached a new level when the steam engine could be carried from site to site.

Courtesy of Young Research Library, UCLA, *Official catalogue of the Great Exhibition of the works of industry of all nations,* London, Spicer Brothers, 1851

higher knowledge, the teaching of natural and experimental philosophy."[98] The marriage between science and industry conceived by Bacon, put into practice by the scientific lecturers of the eighteenth century, and actualized in the factories of men like Watt and Boulton or M'Connel and Kennedy had become the basis of a credo: The union of hand and head make innovation possible.

6a & b. Fine handiwork. The exhibition may have lionized industry, but aristocratic taste prevailed.

Courtesy of Young Research Library, UCLA, *Official catalogue of the Great Exhibition of the works of industry of all nations*, London, Spicer Brothers, 1851

The catalog's proofs and the text itself were written and corrected by a committee of "scientific gentlemen." In some cases the proofs were sent out to the owners of the equipment to make sure that the gentlemen had gotten it right. The spirit of Bacon and Robert Boyle was invoked—the need for the natural philosopher to have insight into the trades. The committee made "an attempt—to convert the changing and inaccurate conventional terms of trade into the precise and enduring expressions of science."[99] Remarkably, the majority of the committee were Fellows of the Royal Society and not industrialists. Clearly the interface between science and manufacturing was sufficiently close at mid-century that the scientifically educated and presumably innovative could understand industrial devices and explain them to the general public. Science worked and, combined with the experience that only hands-on labor could give, made an Industrial Revolution happen. In 1851 the exhibition was used to suggest that the British way of local initiatives and dedication to practical science would forever trump all competitors.

The public responded in droves, and visiting the exhibition became the highlight of any trip to London. Special trains and fares were put on for workers. New encyclopedias of the "useful arts" appeared, directly inspired by the Great Exhibition. Everything from abattoirs to hair pencils was given clear exposition so it could be imitated and improved on.[100] Not least, the seeming infinitude of new gadgets and products fed consumerism. The high level of consumption that made the British market so dynamic in the eighteenth century had become a way of life. Industrial development fueled by invention and knowledge made buying the new endlessly attractive. In the mid-nineteenth century, no other country turned over as much capital per head as did the British.

But others aspired. By the 1820s in America, reformers argued that only progress "in learning and science" would remove the dependence on Europe that the new states still experienced. Education

7. Opening of the Great Industrial Exhibition of All Nations by Her Most Gracious Majesty Queen Victoria and His Royal Highness Prince Albert, on May 1, 1951.

Courtesy of the Grunwald Center for the Graphic Arts, UCLA; photo by Robert Wedemeyer

at home, and not abroad, and the improvement in scientific education should be the first order of business. Following the pattern seen in places like Manchester, aspiring American artisans set up or revitalized institutes for mechanics.[101] Continental governments funneled resources into industry and designed scientific curriculum for the schools so that one day their pupils could manufacture and buy as did the British. The Great Exhibition made the industrial seem to be the order of the world's future, and scientific education seemed to hold the key to success in that future. Competition required the involvement of the state, at least at the level of confidence-building and industry-friendly policies. What the British could not see in 1851 became gradually evident: States could do more than sit on the sidelines, blessing exhibitions for which they did not have to pay, while deferring to industrialists. They could build industries and educational systems that served them. Captured by the exuberant moment of the exhibition, innovation held the key to industrial development. The race belonged to the swift and the scientifically talented—then and now.

Epilogue

By the 1820s Leeds had become like Manchester and Birmingham, a vibrant center of industrial activity. At the Philosophical and Literary Society of the town, lecturers vied to have the chance to make money teaching a course in chemistry or geology, in "the uses of intellectual philosophy," or on the destination of the Nile River in Africa.[1] Atomic theory jostled with the history of ancient Athens for the interest of local merchants and industrialists. While prospering, in their leisure these lecturers were also examining the entire globe, learning its contours and secret riches, acquiring knowledge fit for forging an empire. A few years earlier, and a few hundred miles to the east across the English Channel in Liège, a Belgian professor of mathematics fretted about the absence of a chemistry laboratory at his school. He appealed to the occupying French authorities headquartered back in Paris and reminded them of the usefulness of chemistry and mineralogy for manufacturing.[2] The French wanted the empire they had acquired in 1795 to prosper, to be both industrial and scientific. Partly led by French example, the studied turn toward application with an industrial and imperial focus happened everywhere in the West, and scientific settings were increasingly given an applied and industrial focus.

By 1800 there was barely a place in Western Europe, and even in the newly independent American states, where science escaped valorization. However abstract we may imagine the scientific discipline of the time as being, it was brought down to earth, sometimes literally. The professor of mathematics in Liège taught calculus and trigonometry but also devoted two months to lessons on terrain and the measurement of elevation for use in maps; while his colleague, also in mathematics, taught arithmetic "relative to commerce and to mathematics, the new system of weights and measures," and decimalization. In nearby Ghent the professor of chemistry and experimental physics in the second year of the course taught about the properties of water, about thermometers, optics, theory of colors, and so on, and then paid considerable attention to the metals that appear in mines, the extraction of minerals, the use of specific gravity to identify substances, and to an examination of the principal substances found in the region. He also gave a course particularly for commercial students.[3] By 1820 Ghent held an industrial exposition at which its metal industries figured prominently. Because of its cotton industry, it had become known as Manchester on the Continent.[4]

After 1800 the Napoleonic conquerors inherited the ideals of the French revolutionaries and carried them throughout Europe. They deserve much of the credit for making the ideology of "practical matter" Continental, indeed Western, in its breadth and imperial in its sweep. In the newly occupied territories, from Turin to Amsterdam, men were sought to join the new French implant, *la Société d'encouragement pour l'industrie nationale*.[5] At the same time modernizing Spanish owners of slaves and sugar plantations in Cuba sought to imitate the advances in science and industry witnessed in France and Britain. They argued that "experience" should be "guided by scientific principles."[6] Scientific societies also provided what historians see as social networks where skills learned on the shop floor were passed along. To be sure some technical processes such as dyeing or sugar boiling could only be learned by hands-on experience. But knowing how to work systematically and in disciplined groups could also be learned early and well at the new schools and societies with their technical orientation.[7] In the new scientific culture that matured in the eighteenth century, science bled into technique, and both served the cause of technological innovation.

In this book we have sought to demonstrate the unprecedented expansion of scientific understanding in the many publics to be found in the urban centers of Europe and America. More so than anywhere else, English cities and towns led the way, but European and American audiences and scientific practitioners quickly followed suit. The implications for economic development were momentous. As many economic historians (led by Joel Mokyr) are now having to acknowledge, scientific culture cannot be divorced from the making of early industrialization. Some historians have wanted to confine the Western "takeoff" to the period after 1800, but that is to miss developments at work a good three generations earlier, particularly in England.[8] Scientific culture permitted a shared technical vocabulary to emerge that was understood by engineers and entrepreneurs. Together they worked at mining sites, factories, and canals armed with mechanical knowledge found in Newtonian textbooks, explicated by Newtonian lecturers from Jean Desaguliers working in the 1720s to John Dalton, a Manchester lecturer at work around 1800.

As Francis Bacon well knew, knowledge of nature could also translate into new power. Scientific culture bred an intangible self-confidence, a willingness to put profit ahead of social pedigree, or a desire for universal reform in society and government. By late in the eighteenth century radical politics lurked noticeably in scientific circles, and the guardians of wealth and property were clearly worried. The story we have told about the service to industry and empire rendered by Newton's science includes the challenge to the masters of industry and empire that scientifically informed thinking could also pose. Science did not just allow people to become rich; it also allowed them to dream. If education could change the human understanding of physical nature and teach mere artisans to better use and invent machinery, could science not also suggest progress on every front? Health and wealth could be transformed and made accessible as never before. The democratic challenge to the old order emerged late in the eighteenth century. Its promises remain, only partially fulfilled, but now, like basic science, available to all who have access to a modern education.

NOTES

ACKNOWLEDGMENTS

INDEX

Notes

Introduction

1. L. Osbat, *L'Inquisizione a Napoli: il processo agli ateistic 1688–1697* (Rome: Edizione di storia e letteratura, 1974), p. 16. Cf. Jonathan Israel, *Radical Enlightenment: Philosophy and the Making of Modernity* (Oxford: Oxford University Press, 2001), chap. 35.
2. Amir R. Alexander, *Geometrical Landscapes: The Voyages of Discovery and the Transformation of Mathematical Practice* (Stanford, Calif.: Stanford University Press, 2002), pp. 171–175.
3. The Bakken Library and Museum, Minneapolis, MS Lecture Notes taken by Pruninger on the lectures given by Prof. Voigt (Jena: 1795), unfolioed, but p. 5.
4. Stanley Chapman, ed., *The Autobiography of David Whitehead of Rawtenstall (1790–1865): Cotton Spinner and Merchant* (Settle, England: Helmshore Local History Society, 2001), p. 58.
5. Hewson Clarke and John Dugall, *The Cabinet of Arts, or General Instructor in arts, science, trade, practical machinery . . . and political economy* (London: J. McGowan, 1817), p. 839; cf. [Anon.] *The Artisan; or Mechanic's Instructor . . . geometry, mechanics,* 2 vols. (London: William Cole, 1825); preface preceded by a portrait of Isaac Newton.
6. See [Anon.] *The Artisan*, pp. 90–91, on Thomas Simpson, who became a teacher of mathematics in the evenings at Spitalfields in the 1730s. Eventually he became a Fellow of the Royal Society.
7. [Anon.] *Arcana of science and art, or, One thousand popular inventions and improvements . . . abridged from the transactions of public*

societies, and from the scientific journals . . . illustrated with engravings (London: John Limbird, 1828), pp. 90–92, on domestic economy.

8. J. R. Delaistre, *La science de l'ingénieur* (Lyon: Brunet, 1825), vol. 2, pp. 250–252.

9. A. Quetelet, *Sur l'homme et le développement de ses facultés . . .* (Paris: Bachelier, 1835), vol. 2, p. 251.

1. The Newtonian Revolution

1. For a brief look at the private Newton, see Stephen D. Snobelen, "'God of Gods, and Lord of Lords': The Theology of Isaac Newton's General Scholium to the *Principia*," *Osiris* 16 (2001): 169–208.

2. A. Rupert Hall and Marie Boas Hall, eds., *Unpublished Scientific Papers of Isaac Newton* (Cambridge, England: Cambridge University Press, 1962), pp. 182–213; see also Isaac Newton, *The Principia: Mathematical Principles of Natural Philosophy*, a new translation by I. Bernard Cohen and Anne Whitman, assisted by Julia Budenz, preceded by a guide by I. Bernard Cohen (Berkeley, Calif.: University of California Press, 1999).

3. For further explication, see Cohen, p. 66.

4. Newton, *Principia*, trans. Cohen and Whitman, p. 424.

5. Ibid., p. 429.

6. Ibid., p. 430.

7. Ibid., p. 509.

8. François de Gandt, *Force and Geometry in Newton's Principia* (Princeton, N.J.: Princeton University Press, 1995), p. 268.

9. See Christopher Hill, *Intellectual Origins of the English Revolution* (Oxford: Oxford University Press, 1966), pp. 20–21; and William Eamon, *Science and the Secrets of Nature* (Princeton, N.J.: Princeton University Press, 1994), pp. 304–306.

10. Newton, *Principia*, trans. Cohen and Whitman, pp. 806–807.

11. Ibid.

12. Ibid., p. 809.

13. Ibid., pp. 827–835.

14. Ibid., p. 940.

15. Ibid., p. 943.

16. A modern edition of the work edited by Ezio Vailati can be found:

Samuel Clarke, *A Demonstration of the Being and Attributes of God and Other Writings* (Cambridge, England: Cambridge University Press, 1998). On some of the political implications of Clarke's theology, see Larry Stewart, "Samuel Clarke, Newtonianism and the Factions of Post-Revolutionary England," *Journal of the History of Ideas* 42 (1981): 53–71; Steven Shapin, "Of Gods and Kings: Natural Philosophy and Politics in the Leibniz-Clarke Disputes," *Isis* 72 (1981): 187–215. Note that neither author is an historian of political theory.

17. Clarke [1998 ed.], p. 20.
18. Ibid., p. 36.
19. Ibid., p. 54.
20. Ibid., p. 72.
21. Ibid., p. 150, excerpted from Clarke's *A Discourse concerning the Unchangeable Obligations of Natural Religion and the Truth and Certainty of the Christian Revelation* (London, 1706).
22. Samuel Clarke, *A Discourse Concerning the Unchangeable Obligations of Natural Religion and the Truth and Certainty of the Christian Religion* (London, 1706), pp. 152–153; discussed in greater detail in Margaret C. Jacob, *The Newtonians and the English Revolution, 1689–1720* (Ithaca, N.Y.: Cornell University Press, 1976), pp. 191–192.
23. William Nicolson, Bishop of Carlisle, to Edward Lhwyd, in Lester M. Beattie, *John Arbuthnot: Mathematician and Satirist* (Cambridge, England: Cambridge University Press, 1935), p. 203.
24. Margaret C. Jacob, *Living the Enlightenment: Freemasonry and Politics in Eighteenth-Century Europe* (New York: Oxford University Press, 1991), p. 44.
25. Most easily seen in the English text, Paul Henri Thiry d'Holbach, *The System of Nature: or the Laws of the Moral and Physical World*, vol. 1 (London, 1797), pp. 35–78.

2. The Western Paradigm Decisively Shifts

1. Tsevi Hirsch, *The Book of Rest and Motion (Sefer ha-Menuhah veha Ttenu'ah)* (Vilna, 1867 [originally 1865]). The book was endorsed by the Minister of Public Education in St. Petersburg and the society was thanked for help with publication.
2. Hayyim Zelig Selonimski, *Alexander von Humboldt: The Story of*

His Life, His Voyages and His Books, 2nd ed. (Warsaw: Hayyim Kelter, 1874), p. 9.

3. Ibid., p. 56.

4. Hirsch, *The Book of Rest and Motion,* pp. 299–300.

5. Kenneth J. Howell, *God's Two Books: Copernican Cosmology and Biblical Interpretation in Early Modern Science* (Notre Dame, Ind.: University of Notre Dame Press, 2002).

6. Aristotle, *Physics* 1.1.184a10–25. See also on the complex relationship between Boyle and the Aristotelian tradition, Margaret G. Cook, "Divine Artifice and Natural Mechanism: Robert Boyle's Mechanical Philosophy of Nature," *Osiris* 16 (2001): 133–150; *Science in Theistic Contexts,* eds. John H. Brooke, Margaret J. Osler, and Jitse M. van der Meer.

7. Wiep van Bunge, *From Stevin to Spinoza: An Essay on Philosophy in the Seventeenth-Century Dutch Republic* (Leiden: Brill, 2001), pp. 28–29.

8. David Abercromby, M.D. *Academia Scientiarum: or the Academy of Sciences. Being a Short and Easie Introduction to the Knowledge of the Liberal Arts and Sciences* (London, 1687).

9. Archives de L'Académie royale des sciences, Paris, Procès-Verbaux, vol. 1, Registre de physique, 22 Décembre 1666–Avril 1668, "Project d'Exercitations Physiques, proposé a l'assemblée, par le Sr. Du Clos," f. 3–5.

10. Ibid., p. 110.

11. See L'Académie royale des sciences, Paris, dossier O. Roemer (1644–1710).

12. Allen G. Debus, "The Paracelsians in Eighteenth-Century France: A Renaissance Tradition in the Age of the Enlightenment," pt. 1, *Ambix* 28 (March 1981): 36–54.

13. Rob H. van Gent and Anne C. van Helden, *Een vernuftig geleerde: De technische vondsten van Christiaan Huygens* (Leiden: Museum Boerhaave, 1995); and see also Ellen Tan Drake, *Restless Genius: Robert Hooke and His Earthly Thoughts* (New York: Oxford University Press, 1996).

14. Ephraim Chambers, "Aristotelians," in *Cyclopedia,* vol. 1 (London, 1728), cited in Richard Yeo, *Encyclopaedic Visions: Scientific Dictionaries and Enlightenment Culture* (Cambridge, England: Cambridge University Press, 2001), p. 162.

15. Mr. Morin, *Abregé du Méchanisme universel en discours et questions physiques* (Chartres: Chez J. Roux, 1735), preface.

16. L. W. B. Brockliss, *French Higher Education in the Seventeenth and Eighteenth Centuries* (Oxford: Clarendon Press, 1987), pp. 364–66.

17. M. de Maupertuis, *The Figure of the Earth, Determined from Observations Made by Order of the French King, at the Polar Circle* (London, 1738), pp. 224–225.

18. Ibid., p. 34.

19. Mary Terrall, *The Man Who Flattened the Earth: Maupertuis and the Sciences in the Enlightenment* (Chicago, Ill.: University of Chicago Press, 2002), p. 171.

20. Archives nationales, Paris, Marine G 106, ff. 38–69, 77.

21. David Boyd Haycock, *William Stukeley: Science, Religion and Archaeology in Eighteenth-Century England* (Woodbridge, England: The Boydell Press, 2002), p. 62.

22. Kevin C. Knox, "Lunatick Visions: Prophecy, Signs and Scientific Knowledge in 1790s London," *History of Science* 37 (1999): 432–433.

23. John Martyn and Ephraim Chambers, *The Philosophical History and Memoirs of the Royal Academy of Sciences at Paris* (London, 1742), p. 79.

24. *Verhandelingen uitgegeeven door de Hollandse Maatschappy der Wetenschappen, te Haarlem,* vol. 1 (Haarlem, 1754).

25. Royal Society of London, MSS C. P., item 8, ff. 66–80; on getting a patent, see Christine MacLeod, "Patents for Invention and Technical Change in England, 1660–1753," Ph.D. diss., Cambridge University, 1982, p. 247.

26. Ibid., pp. 12–13.

27. E. Cohen and W. A. T. Cohen-De Meester, "Het Natuurkundig Genootschap der Dames te Middelburg (1785–1887)," *Chemisch Weekblad* 39 (1942): 242–246. See also Claudette Baar-De Weerd, "Het Natuurkundig Genootschap der Dames te Middelburg (1785–1887)," *Zeeland* 10 (2001): 81–90. This essay does not address the larger historiography about women and science. For a more complete discussion of the Middleburg women's society, see Margaret C. Jacob and Dorothée Sturkenboom, "A Women's Scientific Society in the West: The Late Eighteenth-Century Assimilation of Science," *Isis* 33 (June 2003): 217–252.

28. Margaret C. Jacob, *Scientific Culture and the Making of the Industrial West* (New York: Oxford University Press, 1997), pp. 89–92. See also J. W. Buisman, *Tussen Vroomheid en Verlichting: Een cultuurhistorisch en - sociologisch onderzoek naar enkele aspecten van de Verlichting in Nederland (1755–1810)* (Zwolle: Waanders, 1992) and Rienk Vermij, "Science and Belief in Dutch history," in Klaas van Berkel, Albert van Helden, and Lodewijk Palm, eds., *A History of Science in the Netherlands Survey: Themes and Reference* (Leiden, Boston, and Köln: Brill, 1999), pp. 332–347.

29. Christophorus Henricus Didericus Ballot, *Oratio de physices studio, christiano [. . .] quod et ipsae Sacrae testantur Literae, dignissimo eique perituli [. . .]* (Middelburg: Gillissen, 1789); and Herm. Jo. Krom, *Betoog dat de beoefening der natuur- en sterrenkunde niet strijdig is met de erkentenis der godlyke openbaring, en den geopenbaarden godsdienst van Jezus Christus* (Middelburg: Gillissen, 1790). See also Bert Paasman, *J. F. Martinet: Een Zutphens philosooph in de achttiende eeuw* (Zutphen: Van Someren, 1971), p. 28.

30. Whitfield J. Bell, Jr., *Patriot-Improvers: Biographical Sketches of Members of the American Philosophical Society, 1743–1768*, vol. 1 (Philadelphia, Pa.: American Philosophical Society, 1997), pp. 363–64.

31. Ibid., pp. 3–7.

32. Ibid., p. 25, writing in 1773.

33. Anthony Pagden, ed. *The Idea of Europe: From Antiquity to the European Union* (Cambridge, England: Cambridge University Press, 2002), p. 11.

34. Michael D. Gordin, "The Importation of Being Earnest: The Early St. Petersburg Academy of Sciences," *Isis* 91 (2000): 10.

35. George S. Rousseau and David Haycock, "The Jew of Crane Court: Emanuel Mendes da Costa (1717–91)," *History of Science* 38 (2000): 133.

36. For example, see Richard Waller, trans., *Essayes of Natural Experiments Made in the Academie Del Cimento, Under the Protection of the Most Serene Prince Leopold of Tuscany* (London, 1684). For the 1980s vision of Boyle and Big Science, see Steven Shapin and Simon Schaffer, *Leviathan and the Air Pump* (Chicago, Ill.: University of Chicago Press, 1985).

37. James E. McClellan, III, *Science Reorganized: Scientific Societies in*

the Eighteenth Century (New York: Columbia University Press, 1985), p. 160.

38. See Julian Martin, Francis Bacon, the State and the Reform of Natural Philosophy (Cambridge, England: University of Cambridge Press, 1992), pp. 45–56.

39. Charles Webster, The Great Instauration: Science, Medicine and Reform, 1626–1660 (New York, 1975), pp. 387–395.

40. William Eamon, Science and the Secrets of Nature (Princeton, N.J.: Princeton University Press), pp. 335–345.

41. Thomas Sprat, History of the Royal Society (London, 1667), p. 76.

42. Ana Simòes, Ana Carneiro, and Maria Paula Diogo, "Constructing Knowledge: Eighteenth-Century Portugal and the New Sciences," in Kostas Gavroglu, ed. The Sciences in the European Periphery during the Enlightenment. Vol. 2, Archimedes (Dordrecht: Kluwer, 1999), pp. 3–4.

43. A. Nieto-Galan, "The Images of Science in Modern Spain," in Kostas Gavroglu, ed. The Sciences in the European Periphery during the Enlightenment. Vol. 2, Archimedes, pp. 75–77.

44. Ayval Ramati, "Harmony at a Distance: Leibniz's Scientific Academies," Isis 87 (1996): 430–452.

45. Donata Brianto, "Education and Training in the Mining Industry, 1750–1860," Annals of Science 57 (2000): 267–299.

46. Clive Trebilcock, The Industrialization of the Continental Powers, 1780–1914 (London: Longman, 1981), pp. 63–65.

47. For s'Gravesande's admission into the society that published the Journal Litteraire, see University Library, Leiden, Marchand MSS 2, letter from St. Hyacinthe, March 2, 1713. See Marchand MS 1, September 16, 1713, from F. le Bachellé in Utrecht to the society, saying he is afraid of writing for fear of revealing "the secrets of the society."

48. For a longer discussion, see Margaret C. Jacob, The Radical Enlightenment: Pantheists, Freemasons and Republicans, 2nd ed. (Morristown, N.J.: Temple Books, 2003).

49. University Library, Leiden, Marchand MSS 1, September 16, 1713, and December 14, 1713, from F. le Bachellé in Utrecht to the society. In the same collection, 18 9bre 1712, Fritsch in Paris to Marchand on seeing Douxfils in Brussels.

50. John Harris, Lexicon Technicum: Or, An Universal English Dictionary, vol. 2 (London, 1736), p. 1.

51. Ibid., n.p. under "inc."

52. John Harris, *Lexicon Technicum: or an Universal English Dictionary of the Arts and Sciences* (London, 1704), n.p. arranged alphabetically.

53. See John Kersey, *Dictionarium Anglo-Britannecum: Or, A General English Dictionary . . . as also, of all Terms relating to Arts and Sciences, both Liberal and Mechanical,* 2nd ed. (London, 1715), arranged alphabetically.

54. Ibid., for "incidence" (in *Opticks*) the place where two lines or rays meet. For a similar definition, see N. Bailey, *Dictionarium Britannicum: Or a more Compleat Universal Etymological English Dictionary Than any Extant* (London, 1730), arranged alphabetically. Note the absence of the optical meaning of the word (the geometrical is given). A. Boyer, *Le Dictionnaire Royal François-Anglois,* vol. 1 (London, 1773); and see the same and very little science at all in part 1 in A. Boyer, *The Royal Dictionary abridged in Two Parts, French and English.* Pt. 2, *English and French* (London, 1747).

55. On how this was probably invented by the publisher, see Yeo, *Encyclopaedic Visions,* p. 62.

56. J. H., FRS, *Astronomical Dialogues Between a Gentleman and a Lady* (London, 1719), p. 104.

57. Shelley Costa, "The *Ladies' Diary:* Gender, Mathematics, and Civil Society in Early Eighteenth-Century England," in Lynn K. Nyhart and Thomas H. Broman, "Science and Civil Society," *Osiris* 17 (2002): 49–73.

58. Mokyr, *The Gifts of Athena,* p. 67.

59. Yeo, *Encyclopaedic Visions,* p. 151.

60. Pieter van der Star, ed. and trans., *Fahrenheit's Letters to Leibniz and Boerhaave* (Amsterdam: Rodopi, 1983), p. 15.

61. University Library, Leiden, MS BPL 772.

62. F. Macours, "L'enseignement technique à Liège au xviiie siècle," *Bulletin de l'Institute archeologique liegois* 69 (1952): 24–31.

63. Archives departmentales du Nord, MS L 1038, L 155, 295, L 395, 396.

64. Archives nationales, Paris, MS F17 1344 1, cours de physique experimentale, ff. 12–55.

65. Ibid., n.f. from Moulins, 22 fructidor year 6.

66. Archives d'etat Liège, Fonds Français Prefecture, inv. nr. 448 and 458.

67. Archives nationales, Roederer, MS 29 AP 75, f. 395—on the primary schools of England; there is no uniformity, especially in the grammar schools, which as a result of English prosperity have inadver-

tently improved the level of education and have come to form the students for universities; on the subject of "calcul," "C'est cette partie de l'enseignement qui à acquis la plus de perfection chez les Anglais."

68. Ibid., f. 397 directive from Chaptal on seeing to it that Roederer have each *arrondisement* set up its primary school; directive of April 5, 1802 (f. 399) said that mathematics was to be taught in secondary schools and the government would pay for students who secured a place in the *lycée*. There would be thirty *lycées* and Paris would have three. The rest would be in other French cities, but this list included Brussels. Mathematics, pure and applied, was to be taught in all its parts as well as experimental physics, chemistry, natural history, statistics, and technology. There would be professors for physics and chemistry as well as for applied mechanics and technology. The goal was to have 6,000 students in the *lycées*, with 3,000 chosen by the government from the children of functionaries who served the republic well and the other 3,000 chosen by exam. There was a six-year course of study, and the government could distribute its largesse unequally. Eventually La Fleche and one other school were added, and 6,400 pupils became the goal. See f. 645, the professors were to use books approved by the government and the government consulted Delambre and Cuvier at the Institute.

69. Archives nationales, MS F 17 4559, n.f.

70. Joel Mokyr, *The Gifts of Athena: Historical Origins of the Knowledge Economy* (Princeton, N.J.: Princeton University Press, 2002), p. 66.

3. Popular Audiences and Public Experiments

1. James Van Horn Melton, *The Rise of the Public in Enlightenment Europe* (Cambridge, England: Cambridge University Press, 2001), pp. 11–13, 106–107.

2. Larry Stewart, *The Rise of Public Science* (Cambridge, England: Cambridge University Press, 1992); Jan Golinski, *Science As Public Culture: Chemistry and Enlightenment in Britain, 1760–1820* (Cambridge, England: Cambridge University Press, 1992); Patricia Fara, *Sympathetic Attractions: Magnetic Practices, Beliefs, and Symbolism in Eighteenth-Century England* (Princeton, N.J.: Princeton University Press, 1996), esp. p. 37.

3. Simon Schaffer, "Enlightened Automata," in William Clark, Jan Golinski, and Simon Schaffer, eds., *The Sciences in Enlightened Europe* (Cambridge, England: Cambridge University Press, 1999), pp. 126, 129; Schaffer, "Machine Philosophy: Demonstration Devices in Georgian Mechanics," in Albert Van Helden and Thomas L. Hankins, eds., "Instruments," *Osiris* 9 (1993): 157–182.

4. On the role of the gentleman and status in asserting discovery, see Steven Shapin, *A Social History of Truth: Civility and Science in Seventeenth-Century England* (Chicago and London: University of Chicago Press, 1994).

5. Clark, Golinski, and Schaffer, eds., *The Sciences in Enlightened Europe*, p. 23.

6. Quoted from the Wedgwood manuscripts in Robin Reilly, *Josiah Wedgwood 1730–1795* (London: Macmillan, 1992), pp. 18–19.

7. On Whiston's career, see Stewart, *The Rise of Public Science*, esp. pp. 94 ff.; and Larry Stewart and Steven Snobelen, "Making Newton easy: William Whiston in Cambridge and London," in Kevin C. Knox and Richard Noakes, eds., *From Newton to Hawking. A History of Cambridge University's Lucasian Professors of Mathematics* (Cambridge: Cambridge University Press, 2003), 135–170. On the role of coffeehouses, see Melton, *The Rise of the Public in Enlightenment Europe*, pp. 240–249.

8. Fara, *Sympathetic Attractions*, pp. 39 ff.

9. William Whiston, *Memoirs of the Life and Writings of Mr. William Whiston* (London, 1749), pp. 249–251; See also James E. Force, *William Whiston: Honest Newtonian* (Cambridge, England: Cambridge University Press, 1985), pp. 23–24.

10. See Simon Schaffer, "The Consuming Flame: Electrical Showmen and Tory Mystics in the World of Goods," in John Brewer and Roy Porter, eds., *Consumption and the World of Goods* (London and New York: Routledge, 1993), pp. 489–526; and Larry Stewart, "Seeing Through the Scholium: Religion and Reading Newton in the Eighteenth Century," *History of Science* 34 (1996): 123–165, esp. 151.

11. Robert M. Isherwood, *Popular Entertainment in Eighteenth-Century Paris* (Oxford: Oxford University Press, 1986), pp. 48–50.

12. Schaffer, "Enlightened Automata," p. 135; Stewart, *The Rise of Public Science*, pp. 123–127.

13. John Harris, *Lexicon Technicum Or An Universal English Dictionary of Arts and Sciences,* vol. 1 (London, 1704; New York and London:

Johnson Reprint, 1966), see "Engine."

14. See Thomas L. Hankins, *Science and the Enlightenment* (Cambridge and New York: Cambridge University Press, 1985), pp. 48–50; also Stewart, *The Rise of Public Science* (1992), passim.

15. Vaucanson to Daniel Charles Trudaine, 1765; quoted in Robin Briggs, "The Academie Royale des Sciences and the Pursuit of Utility," *Past and Present* 131 (May 1991): 84.

16. See Alan Morton, "Concepts of Power: Natural Philosophy and the Uses of Machines in Mid-Eighteenth Century London," *British Journal for the History of Science* 28 (March 1995): 63–78, esp. 73–77; Alan Q. Morton and Jane Wess, *Public and Private Science: The King George III Collection* (Oxford: Oxford University Press, 1993), chap. 4; and Schaffer, "Enlightened Automata," pp. 145–147.

17. See the excellent popular account by Stephen Pumfrey, *Latitude and the Magnetic Earth* (Duxford: Icon Press, 2002), esp. p. 4; and Lisa Jardine, *Ingenious Pursuits: Building the Scientific Revolution* (New York and London: Talese and Doubleday), chap. 4.

18. Newton, possibly to the Admiralty, n.d. Cambridge University Library, Add. MSS. 3972, f. 37. On Whiston and the longitude, see Stewart, *Rise of Public Science,* pp. 186 ff.

19. See Simon Schaffer, "Glass Works: Newton's Prisms and the Uses of Experiment," in David Gooding, Trevor Pinch, and Simon Schaffer, eds., *The Uses of Experiment: Studies in the Natural Sciences* (Cambridge, England: Cambridge University Press, 1989), pp. 67–104, esp. 95–96.

20. [William Whiston], *Several Papers Relating to Mr. Whiston's Cause Before the Court of Delegates* (London, 1715), pp. 4, 15, 24.

21. See J. L. Heilbron, *Physics at the Royal Society during Newton's Presidency* (Los Angeles: William Andrews Clark Memorial Library, 1983), and Larry Stewart, "Other Centres of Calculation, or, Where the Royal Society Didn't Count: Commerce, Coffee-houses and Natural Philosophy in Early Modern London," *British Journal for the History of Science* 32 (1999): 133–153.

22. William Eamon, *Science and the Secrets of Nature: Books of Secrets in Medieval and Early Modern Culture* (Princeton, N.J.: Princeton University Press, 1994), pp. 309–310.

23. The Bakken Library and Museum, MS, Lectures by Johann Heinrich Voigt, Jena, 1795, n.f.

24. The Bakken Library and Museum, MS, "Lectures on Experimental

Philosophy delivered at the Medical Theatre of Guy's Hospital, 1808 by William Allen, & Lectures on Chemistry at the same hospital and on Natural Philosophy at the Royal Institution. Taken in notes at the Lecture Rooms." The book is signed by J. Couch.

25. The Bakken Library, MS, "Cahier des physique experimentale . . . par M. Sartre," given at the Ecole centrale at Laval, n.f. section on "Des propriétés générales de la matières."

26. Francis Hauksbee, *Esperienze fisico-meccaniche sopra varj soggetti contenenti un racconto di diversi stupendi fenomeni* (Florence, 1716).

27. For Nollet, see *Programme, ou, Idée générale d'un cours de physique* (Paris, 1738); Joh. Henrici Mulleri, *Collegium experimentale . . . de Aere, Aqua, Igne ac Terrestribus* (Nuremberg, 1721). Cf. Pierre Poliniere, *Experiences de physique* (Paris, 1709).

28. Voltaire, *Letters on England,* trans. Leonard Tanock (Harmondsworth, England: Penguin, 1980), p. 71.

29. The best single account of the rise of electricity remains J.L. Heilbron, *Electricity in the 17th and 18th Centuries: A Study of Early Modern Physics* (1979; Mineola, N.Y.: Dover, 1999).

30. Benjamin Franklin, *Experiments and observations on electricity, made at Philadelphia in America* (London: E. Cave, 1751), preface.

31. Louis de Bougainville, *Voyage autour du monde, Par La Fregate du Roi, La Boudeuse, et La Flute L'Etoile; En 1766, 1767, 1768 & 1769* (Paris: Chez Saillant & Nyon, 1771), pp. 64, 383.

32. See, for example, *Recueil de traité sur l'electricité, Traduits de l'Allemand & de l' Anglois* (Paris, 1748), with treatises by F. Winckler at the University of Leipzic and William Watson of the Royal Society of London.

33. Franklin, *Experiments and Observations,* p. 49.

34. Ibid., pp. 87–88; see also Jessica Riskin, *Science in the Age of Sensibility: The Sentimental Empiricist of the French Enlightenment* (Chicago, Ill.: University of Chicago Press, 2002), esp. chaps. 3–4.

35. Giuliano Pancaldi, *Volta: Science and Culture in the Age of Enlightenment* (Princeton, N.J.: Princeton University Press, 2003), pp. 80–84.

36. For the background to collecting and the growth of cabinets, see Paula Findlen, *Possessing Nature: Museums, Collecting, and Scientific Culture in Early Modern Italy* (Berkeley, Calif.: University of California Press, 1996; Lisa Jardine, *Ingenious Pursuits*; and John V. Pickstone, *Ways of Knowing: A New History of Science, Technol-*

ogy and Medicine (Chicago, Ill.: University of Chicago Press, 2001), pp. 64, 87.

37. Fara, *Sympathetic Attractions,* pp. 118 ff.
38. Geoffrey V. Sutton, *Science for a Polite Society: Gender, Culture, and the Demonstration of Enlightenment* (Boulder, Co.: Westview, 1995), pp. 214–240; Isherwood, *Farce and Fantasy,* p. 49; and Barbara Maria Stafford, *Artful Science: Enlightenment Entertainment and the Eclipse of Visual Education* (Cambridge, Mass., M. I. T. Press, 1994), pp. 149–153, 173.
39. See Shelly Costa, "Marketing Mathematics in Early Eighteenth-Century England: Henry Beighton, Certainty, and the Public Sphere," *History of Science* 40 (June 2002): 211–232.
40. Jan Golinski, "Barometers of Change: Meteorological Instruments as Machines of Enlightenment," in Clark, Golinski, and Schaffer, eds., *The Sciences in Enlightened Europe,* pp. 68–93, esp. 83.

4. Practicality and the Radicalism of Experiment

1. John Conduitt, "Newton's Manual Dexterity," King's College, Cambridge, England, MSS. 130 (9). 3–4.
2. See Larry Stewart and Paul Weindling, "Philosophical Threads: Natural Philosophy and Public Experiment Among the Weavers of Spitalfields," *British Journal for the History of Science* 28 (1995): 37–62.
3. Eamon, *Science and the Secrets of Nature,* p. 310.
4. John Grundy, *Chester Navigation consider'd* (n.d., ca. 1736).
5. Jean Desaguliers, *A Course of Experimental Philosophy,* 2nd ed., vol. 1 (London, 1745), pp. 70, 138.
6. Royal Society MSS, Certificates, 2 (1751–1756).
7. John Smeaton, "An Experimental Examination of the Quantity and Proportion of Mechanic Power Necessary to Be Employed in Giving Different Degrees of Velocity to Heavy Bodies from a State of Rest," *Philosophical Transactions of the Royal Society* 46 (London, 1777).
8. Cf. Clark, Golinski, and Schaffer, eds., *The Sciences in Enlightened Europe*; and Lorraine Daston, "Enlightenment Calculations," *Critical Inquiry* 21 (Autumn 1994): 182–202.
9. Paul Langford, *Englishness Identified: Manners and Character 1650–1850* (Oxford: Oxford University Press, 2000), p. 76.

10. Henry Beighton, *The Ladies Diary: or the Woman's Almanack* (London, 1721).

11. Birmingham Central Library, Matthew Boulton Papers 254. Clement Smith, Richmond Water Works, to Boulton and Watt, August 29, 1778.

12. H. W. Dickinson and Rhys Jenkins, *James Watt and the Steam Engine*, 2nd ed. (London: Moorland, 1981), pp. 353–355; cf. Jenny Uglow, *The Lunar Men: Five Friends Whose Curiosity Changed the World* (New York: Farrar, Straus & Giroux, 2002), p. 376.

13. Quoted in Dickinson and Jenkins, *James Watt and the Steam Engine*, pp. 355–356.

14. See Paola Bertucci, "The Electrical Body of Knowledge: Medical Electricity and Experimental Philosophy in the Mid-Eighteenth Century," in Paola Bertucci and Giuliano Pancaldi, eds. *Electric Bodies: Episodes in the History of Medical Electricity* (Bologna: CIS, Dipartimento di Filosofia, 2001), pp. 43–68, esp. 54.

15. Mary Terrall, *The Man Who Flattened the Earth: Maupertuis and the Sciences in the Enlightenment* (Chicago, Ill.: University of Chicago Press, 2002), pp. 136 ff.

16. Louis de Bougainville, *Voyage autour du monde, Par La Fregate du Roi, La Boudeuse, et La Flute L'Etoile; En 1766, 1767, 1768 & 1769* (Paris: Chez Saillant & Nyon, 1771), pp. 16–17.

17. J. L. Heilbron, *Electricity in the 17th and 18th Centuries: A Study of Early Modern Physics* (Berkeley and Los Angeles: University of California Press, 1979), pp. 380 ff.; Trent A. Mitchell, "The Politics of Experiment in the Eighteenth Century: The Pursuit of Audience and the Manipulation of Consensus in the Debate over Lightning Rods," *Eighteenth-Century Studies* 32 (1998): 307–331; Jean-Pierre Poirier, *Lavoisier: Chemist, Biologist, Economist,* trans. Rebecca Balinski (Philadelphia, Pa.: University of Pennsylvania Press, 1998), p. 150.

18. See William H. Sewell, "Visions of Labour: Illustrations of the Mechanical Arts before, in, and after Diderot's *Encyclopedie,*" in Steven Laurence Kaplan and Cynthia J. Keopp, eds. *Work in France: Representations, Meaning, Organization, and Practice* (Ithaca, N.Y. and London: Cornell University Press, 1986), pp. 258–286, esp. 275.

19. See Simon Schaffer, "Measuring Virtue: Eudiometry, Enlightenment and Pneumatic Medicine," in Andrew Cunningham and Roger French, eds., *The Medical Enlightenment of the Eighteenth Century* (Cambridge, England: Cambridge University Press, 1990), pp. 281–

318; James Stirling, the younger, "A Journal of Travels," p. 33. Scottish Record Office. Stirling of Garden MSS. Bundle 40.

20. Birmingham Central Library, James Watt Papers 4/48/7. Percival to James Watt, September 16, 1786; on Percival, see W. V. Farrar, Kathleen R. Farrar, and E. L. Scott, "Thomas Henry (1734–1816)," in Wilfred Vernon Farrar, *Chemistry and the Chemical Industry in the 19th Century: The Henrys of Manchester and other Studies*, eds. Richard L. Hills and W. H. Brock (Brookfield, Vt.: Variorum, 1997), chap. 1, *passim*.

21. Birmingham Central Library, James Watt Papers W/9/56. Robert Cleghorn to Watt, May 12, 1796; Robert Cleghorn taught chemistry to Gregory Watt in Glasgow and he was one of the minority of radical professors who supported the French Revolution. See Margaret Jacob and Lynn Hunt, "The Affective Revolution in 1790s Britain," *Eighteenth-Century Studies* 34 (2001): 491–521.

22. Christopher Lawrence, "Medical Minds, Surgical Bodies: Corporeality and the Doctors," in Christopher Lawrence and Steven Shapin, eds., *Science Incarnate: Historical Embodiments of Natural Knowledge* (Chicago, Ill.: University of Chicago Press, 1997), pp. 156–201.

23. Cornwall Record Office, Davies Gilbert correspondence. DG41/54, Thomas Beddoes to Davies Gilbert (Giddy), October 8, 1792.

24. Birmingham Central Library, James Watt Papers 4/65/14. Thomas Henry to Watt, April 16, 1795.

25. Birmingham Central Library, James Watt Papers, 4/65/4. Beddoes to Watt, May 29, 1799. On Davy, see Golinski, *Science as Public Culture*, esp. pp. 166–167.

26. A. E. Musson and Eric Robinson, *Science and Technology in the Industrial Revolution* (Toronto: University of Toronto Press, 1969).

27. Quoted in Isaac Kramnick, "Eighteenth-Century Science and Radical Social Theory: The Case of Joseph Priestley's Scientific Liberalism," *Journal of British Studies* 25 (January 1986): 1–30, esp. p. 8.

28. Keir, *The First Part of a Dictionary of Chemistry*, quoted in Golinski, *Science as Public Culture*, p. 147.

29. Joseph Priestley, *Experiments and Observations on Air* (Birmingham, 1790), quoted in Maurice Crosland, "The Image of Science As a Threat: Burke versus Priestley and the 'Philosophic Revolution,'" *British Journal for the History of Science* 20 (July 1987): 277–307, esp. 282.

30. John Conduitt, "Newton's Character," King's College, Cambridge,

MSS. 130 (7). 2; "Account of Newton's Life at Cambridge," typescript. King's College, Cambridge, MSS. 130 (4). 16.

31. George Horne, *A Fair, Candid, and Impartial State of the Case Between Sir Isaac Newton and Mr. Hutchinson. In Which Is Shewn, How far a system of PHYSICS is capable of MATHEMATICAL DEMONSTRATION; how far Sir Isaac's, as such a system, has that DEMONSTRATION; and consequently, what regard Mr. HUTCHINSON'S claim may deserve to have paid it* (Oxford: Printed at the Theatre for S. Parker, 1753), pp. 42, 46.

32. Horne, *A Fair, Candid, and Impartial State of the Case*, p. 55.

33. Burke, *Letter to Noble Lord,* quoted in Crosland, "The Image of Science as a Threat," p. 295; Burke, *Reflections on the Revolution in France,* ed. J. C. D. Clarke (Stanford, Calif.: Stanford University Press, 2001), pp. 240–241; William Jones, *Memoirs of the Life, Studies, and Writings of the Right Reverend George Horne, D. D., Late Bishop of Norwich* (London, 1795), pp. 31–32 & n.

34. Quoted in Dan Eshet, "Rereading Priestley: Science at the Intersection of Theology and Politics," *History of Science* 39 (June 2001): 127–159, esp. 139.

35. See Golinski, *Science as Public Culture,* p. 186.

36. Henry H. Cawthorne, "The Spitalfields Mathematical Society (1717–1845," *Journal of Adult Education* 3 (1929): 158.

37. Heilbron, *Electricity in the 17th and 18th Centuries,* pp. 373 ff.

38. See Stewart and Weindling, "Philosophical Threads: Natural Philosophy and Public Experiment among the Weavers of Spitalfields," *British Journal for the History of Science* 28 (1995): 37–62.

39. *Articles of the Mathematical Society, Spitalfields London: Instituted 1717* (London: 1793), p. 11.

40. See the biography of John Williams in Royal Astronomical Society, *Monthly Notices* 35 (1875): 180–183.

41. *A Catalogue of Books Belonging to the Mathematical Society, Crispin Street, Spitalfields* (London: James Whiting, 1804), pp. ii–iii.

42. Albert Goodwin, *Friends of Liberty: The English Democratic Movement in the Age of the French Revolution* (Cambridge, Mass.: Harvard University Press, 1979), pp. 387–388; Paul Weindling, "Science and Sedition: How Effective Were the Acts Licensing Lectures and Meetings, 1795–1819?," *British Journal for the History of Science* 13 (1980): 139–153.

43. Quoted in Weindling, "Science and Sedition," p. 143.

44. J. W. S. Cassels, "The Spitalfields Mathematical Society," *Bulletin of the London Mathematical Society* 2 (October 1979): 24; Cawthorne, "The Spitalfields Mathematical Society (1717–1845)," 160; Greater London Record Office, MSS. MR/SL/2, 1817.

45. Robert E. Schofield, *The Lunar Society of Birmingham: A Social History of Provincial Science and Industry in Eighteenth-Century England* (Oxford: Clarendon Press, 1963); and Jenny Uglow, *The Lunar Men: Five Friends Whose Curiosity Changed the World* (New York: Farrar, Straus & Giroux, 2002).

46. See Musson and Robinson, *Science and Technology in the Industrial Revolution*, pp. 105–107.

47. See L. Stewart, "Putting on Airs: Science, Medicine and Polity in the Late Eighteenth-Century," in Trevor Levere and Gerard L'E. Turner, eds., *Discussing Chemistry and Steam: The Minutes of a Coffee House Philosophical Society 1780–1787* (Oxford: Oxford University Press, 2002), pp. 207–255, esp. 230–231; and see also J. N. Hays, "The London Lecturing Empire, 1880–1850," in Ian Inkster and Jack Morrell, eds., *Metropolis and Province: Science in British Culture 1780–1850* (Philadelphia, Pa.: University of Pennsylvania Press, 1983), pp. 91–119.

48. Arthur Young, *Political Essays concerning the Present State of the British Empire* (London, 1772; reprint, New York: Research Reprints, 1970), pp. 213, 219.

49. See J. V. Golinski, "Utility and Audience in Eighteenth-Century Chemistry: Case Studies of William Cullen and Joseph Priestley," *British Journal for the History of Science* 21 (March 1988): 1–31.

50. Greater London Record Office, Howard Papers, Acc. 1017/1323. William to Luke Howard, August 15, 1792.

51. See Iwan Rhys Morus, *Frankenstein's Children: Electricity, Exhibition, and Experiment in Early-Nineteenth-Century London* (Princeton, N.J.: Princeton University Press, 1998), p. 14.

52. Samuel Parkes, *Chemical Essays, Principally Relating to the Arts and Manufactures of the British Dominions*, vol. 5 (London: Baldwin, Cradock and Joy, 1815), p. v.

53. Watson, *Chemical Essays*, vol. 2, pp. 39–40; Musson and Robinson, *Science and Technology in the Industrial Revolution*, pp. 167–170.

54. Cf. Robert Fox, "Diversity and Diffusion: The Transfer of Technol-

ogies in the Industrial Age," *Transactions of the Newcomen Society* 70 (1998–1999): 185–196, esp. 186.

55. UCLA, Young Research Library, MS "Lectures in Chemistry by Doctor Black and Doctor Hope," notes taken by Lovell Edgeworth, vol.1, 1796, fols. 83–85, 89–91.

56. Quoted in J. B. Morrell, "Wissenschft in Worstedopolis: Public Science in Bradford, 1800–1850," *British Journal for the History of Science* 18 (March, 1985): 1–23, esp. 11; and Ian Inkster, "The Social Context of an Educational Movement: A Revisionist Approach to the English Mechanics' Institutes, 1820–185," *Oxford Review of Education* 2 (1976): 277–307; Inkster, "The Public Lecture as an Instrument of Science Education for Adults: The Case of Great Britain, c. 1750–1850," *Paedogogica Historica* 20 (1980): 80–107.

5. Putting Science to Work: European Strategies

1. For an excellent sense of what he taught, see UCLA, Young Research Library, MS "Lectures in Chemistry by Doctor Black and Doctor Hope. Taken by Lovell Edgeworth . . . Edinburgh, 1796."

2. Paul Wood, intro. *Essays and Observation, Physical and Literary. Read before a Society in Edinburgh,* vols 1–3 (Bristol, England: Thoemmes, 2002.

3. [J. Palairet], *Abregé sur les sciences & sur les arts . . . A short Treatise upon Arts and Sciences* (London, 1731), a French textbook in mathematics and literature used in England also to teach French and suitable for children in the later stages of a grammar school. See also R. V. and P. J. Wallis, *Biobibliography of British Mathematics and its Applications, Part II: 1701–1760* (Newcastle upon Tyne: Epsilon Press, 1986).

4. Philippe Minard, "Colbertism Continued? The Inspectorate of Manufactures and Strategies of Exchange in Eighteenth-Century France," *French Historical Studies* 23 (2000): 477–496.

5. Rachel Lauden, "Cognitive Change in Technology and Science," in R. Lauden, ed. *The Nature of Technological Knowledge: Are Models of Scientific Change Relevant?* (Boston, Mass.: D. Reidel Publishing, 1984), p. 92.

6. The book was Pierre Sigorgne, *Institutions Newtoniennes* (Paris, 1747).

7. This story is summarized from Helena M. Pycior, *Symbols, Impossible Numbers, and Geometric Entanglements: British Algebra through the Commentaries on Newton's Universal Arithmetick* (Cambridge, England: Cambridge University Press, 1997), chap. 7.

8. Cited in Pycior, p. 248.

9. George Berkeley, *The Analyst*, in George Sampson, ed., *The Works of George Berkeley, D. D., Bishop of Cloyne*, vol. 3 (London: George Bell, 1898), p. 33.

10. By the author of *The Minute Philosopher* [Bishop Berkeley], *A Defence of Free-Thinking in Mathematics* (London, 1735), p. 6.

11. Philalethes Cantabrigiensis [James Jurin], *Geometry no Friend to Infidelity: or, a Defence of Sir Isaac Newton and the British Mathematicians* (London, 1734), p. 9; for the comment about the Spanish Inquisitor, see p. 27.

12. Ibid., pp. 9–10.

13. Ibid., p. 49.

14. Fabio Bevilacqua and Lucio Fregonese, eds. *Nuova Voltiana: Studies on Volta and His Times*, vol. 1 (Milan: Editore Ulrico Hoepli, 2000), pp. 64–70.

15. Sadi Carnot, *Reflections on the Motive Power of Fire*, ed. with an introduction by E. Mendoza (New York: Dover Publications, 1960).

16. Michel Cotte, "La circulation de l'information technique, une donnée essentielle de l'initiative industrielle sous la Restauration," in André Guillerme, ed., *De la diffusion des sciences à l'espionnage industriel XV e–XX e siècle: Actes du colloque de Lyon (30–31 mai 1996).* (Paris: Ecole normale superiore, 1999), pp. 133–158.

17. François Crouzet, *The First Industrialists* (Cambridge, England: Cambridge University Press, 1985), chap. 1.

18. Maria and R. L. Edgeworth, *Practical Education*, 3 vols. (London, 1801; reprint, Poole, England: Woodstock Books, 1996); in particular, volume 2, largely written by Richard. Cf. on what needed to be known by workers: *Report of the Committee of the Birmingham Mechanics' Institution, Read at the Ninth Anniversary Meeting, Held Friday, January 2, 1835, in the Lecture Room, Cannon-Street* (Birmingham: Printed by J. W. Showell, 1835).

19. For extensive information on M'Connel, see the memoir by his son, David C. M'Connel, *Facts and Traditions Collected for a Family Record* (Edinburgh: Printed by Ballantine and Co. for private circula-

tion, 1861). (Found in Manchester Central Library, Social Science Reference, Q929.2 M76)

20. *Official Descriptive and Illustrated Catalogue by Authority of the Royal Commission,* vol. 1 (London: Spicer Bro., 1851), pp. xxiv–xxv; Manchester had 191 exhibits and Birmingham had 230.

21. MCK/2/2/2, to Boulton & Watt, July 1, 1797, responding to letter of May 26. "It is recommended to us by some of our experienced friends to have the boiler made larger and stronger than you commonly do for that power; Although it costs more we mention this that you may not be limited, when there appears to be an advantage"; to Boulton and Watt, October 5, 1798: "We have got the Cylinder and base here today and are now very much in want of cement to put them together with. Please to Forward a Box of it by the first Coach if possible— have likewise Got a Beam"; to Boulton and Watt, October 17, 1798: "We find the Planet wheel is so much damaged that [it] may break when the Engine is Set to work. Therefore please to send one as soon as possible with the pin fitted to it that was ordered for the Double Link"; to Boulton and Watt, January 16, 1799: "Having had some conversation with some of the Partners in the Underwood Spinning Comp., Paisley Respecting our Cylinder & Piston &c. that lay here which they have no objection to take for their Engine if you think they are as good as new. Shall be very Glad if you can bring it in the Price we leave to you. We believe the[y] are as good as can be made"; to Boulton and Watt, June 21: 1802: "Requesting B&W to send as soon as possible . . . the "Crank & Shafts & Fly Wheel Shaft with the wheels belong'g as our millwrights are nearly at a stand for want of them."

22. MCK/2/1/10 Letters Received. John Southern for Boulton & Watt to M'Connel & Kennedy, November 30, 1804. A longer version of this section about Manchester first appeared in the *Canadian Journal of History* 36 (2001): 283–304, coauthored by Margaret Jacob and David Reid.

23. MCK/2/2/2 Letters Sent, June 2, 1796–June 14, 1805; [103] to Boulton and Watt, October 17, 1798.

24. Iwan Rhys Morus, *Frankenstein's Children: Electricity, Exhibition, and Experiment in Early Nineteenth-Century London* (Princeton, N.J.: Princeton University Press, 1998), p. 189.

25. Derek Fraser, ed., *Municipal Reform and the Industrial City* (Bath: Leicester University Press, 1982), p. 23.

26. For biographical background on James M'Connel and John Kennedy, see M'Connel, *Facts and Traditions*; John Kennedy, "Brief Notice of My Early Recollections, in a Letter to My Children," in idem, *Miscellaneous Papers on Subjects Connected with the Manufactures of Lancashire* (Manchester: For private distribution, 1849), pp. 1–18; William Fairbairn, *A Brief Memoir of the Late John Kennedy, Esq.* (Manchester: Charles Simms and Co., 1861); and C. H. Lee, *A Cotton Enterprise, 1795–1840: A History of M'Connel & Kennedy, Fine Cotton Spinners* (Manchester: Manchester University Press, 1972), esp. chap. 2.

27. This is the approach taken throughout in Peter Mathias, "Who Unbound Prometheus?" in Peter Mathias, ed., *Science and Society 1600–1900* (Cambridge, England: Cambridge University Press, 1972).

28. Edward W. Stevens, Jr., *The Grammar of the Machine: Technical Literacy and Early Industrial Expansion in the United States* (New Haven, Conn.: Yale University Press, 1995).

29. Stevens, *Grammar of the Machine*, pp. 2–4. On the pedagogical and epistemological problems associated with graphical representation, see Chapter 2. Although Stevens focuses on the United States, the issues he discusses closely parallel those being considered in Great Britain at the time.

30. Kennedy, "Brief Notice," pp. 4–5.

31. Ibid., p. 14; see also A. E. Musson and Eric Robinson, *Science and Technology in the Industrial Revolution* (Toronto: University of Toronto Press, 1969), p. 108.

32. Kennedy, "Brief Notice," p. 6.

33. See the breadth of topics covered in Kennedy, *Miscellaneous Essays*, which includes articles on manufacturing, the poor laws, and the effect of technology on the working classes.

34. From time to time, membership lists for the Lit and Phil appeared in the *Memoirs of the Literary and Philosophical Society of Manchester*. Dates of election are given in volume 6 of the second series (1842); but by this time, only Kennedy (elected 1803) was still alive. M'Connel's name first appears in the list in volume 2 (1813), but we can assume that he joined the society soon after Kennedy. James

M'Connel, Jr., and his brother William were elected in 1829 and 1838, respectively. Interestingly, just before his death, Kennedy was the oldest living member of the Society. Fairbairn, *Brief Memoir,* p. 10.

35. MCK/2/1/8/3 Printed circular from J. B. Stedman, secretary, to the Board of the Manchester Infirmary and Lunatic Hospital, November 19, 1803. John Rylands Library, Deansgate.

36. M'Connel, *Facts and Traditions,* p. 148. Also see the lists of board members printed in the annual *Report of the Directors of the Manchester Mechanics' Institution.*

37. For Kennedy's involvement in the Manchester/Liverpool railway, see Carlson, *Liverpool and Manchester Railway Project,* pp. 50, 62, 218–219.

38. Wilfred Vernon Farrar, *Chemistry and the Chemical Industry in the Nineteenth Century: The Henrys of Manchester and Other Studies,* eds. Richard L. Hills and W. H. Brock (Aldershot, England: Variorum, 1997), pp. 187–191.

39. For the Society's membership between 1781 and 1852, see Thackray, "Natural Knowledge," table on p. 695.

40. Eaton Hodgkinson, "Some Account of the Late Mr. Ewart's Paper on the Measure of Moving Forces; and on the Recent Applications of the Principles of Living Forces to Estimate the Effects of Machines and Movers," *Memoirs of the Literary and Philosophical Society of Manchester,* 2nd series, 7 (1846): 137–156. On Ewart at Leeds, see the Gott MSS at the Brotherton Library, University of Leeds, MS 193/ 2, f. 38.

41. On the College of Arts and Sciences, see Thomas Barnes's articles in the first two volumes of *Memoirs of the Literary and Philosophical Society of Manchester.* On the relationship between the Lit and Phil and the Dissenting academy known as Manchester College (now called Harris Manchester College, Oxford), see Jean Raymond and John V. Pickstone, "The Natural Sciences and the Learning of the English Unitarians," in Barbara Smith, ed., *Truth, Liberty, Religion: Essays Celebrating Two Hundred Years of Manchester College* (Oxford: Manchester College Oxford, 1986), pp. 127–164, pp. 134–15.

42. In 1849 these four papers were collected and published under the title *Miscellaneous Papers on Subjects Connected with the Manufactures*

of Lancashire. Several were also published as individual pamphlets. A fifth paper was not delivered at the Lit and Phil, but was published separately: John Kennedy, *On the Exportation of Machinery: A Letter Addressed to the Hon. E. G. Stanley, M. P.* (London: Longman, Hurst & Co. et al., 1824).

43. John Kennedy, *Observations on the Rise and Progress of the Cotton Trade in Great Britain, Particularly in Lancashire and the adjoining Counties* (Manchester: The Executors of the Late S. Russell, 1818), p. 20.

44. Kennedy, *Observations*, p. 3.

45. Kennedy, *Observations*, pp. 17–18.

46. MCK/2/2/5 Letters Sent. See, in particular, M'Connel and Kennedy to John Bell & Co., Belfast, February 13, 1816, and March 29, 1816; idem to Robert McGavind & Co., March 28, 1816; idem to W. Sangford, May 23, 1816.

47. M'Connel and Kennedy to John Bell & Co., February 13, 1816.

48. For a general account of the Mechanics' Institutes in Britain, see Ian Inkster, "The Social Context of an Educational Movement: A Revisionist Approach to the English Mechanics' Institutes, 1820–1850," in idem, *Scientific Culture and Urbanisation in Industrialising Britain* (Aldershot, England: Variorum, 1997).

49. On M'Connel's monetary donation, see M'Connel, *Facts and Traditions*, p. 148.

50. *Report of the Directors of the Manchester Mechanics' Institution, May 1828, with the Rules and Regulations of the Institution* (Manchester: Printed by R. Robinson, St. Ann's Place, 1828), p. 23.

51. Ibid., p. 9.

52. See, for instance, the reports for 1828 and 1834.

53. Jeffrey A. Auerbach, *The Great Exhibition of 1851: A Nation on Display* (New Haven, Conn., Yale University Press, 1999), pp. 10–12.

54. A discussion with John Smeaton; see Mike Clarke, *The Leeds and Liverpool Canal: A History and Guide* (Preston, England: Carnegie Press, 1990), p. 70.

55. Brotherton Library, University of Leeds, Gott MS 193/3/f. 98. Letter of Davison to Gott asking Gott if he would go with Davison to give his opinion of their steam engine to Mr. Goodwin: "but if you can't here are queries in writing." May 5, 1802. On the engine and its many

uses for scribbling, carding, turning shafts and gearings, and grinding stones, see H. Heaton, "Benjamin Gott and the Industrial Revolution in Yorkshire," *The Economic History Review* 3 (1931–1932): 52–53.

56. Brotherton, MS 193/ 3 f. 94.

57. Ibid., f. 97 Gott to Bramah from Leeds, March 29, 1809, on his hydro-mechanical press: "We have from your letter of the 25th instant that the sale and general adoption of your patent presses have been prevented by unfavorable representations respecting the merits & utility of the one you erected for us . . . we must . . . tell you that we look after every operation of the work ourselves, and if we had experienced any advantage from the use of your press, we should have insisted on those men working it, or we should have appointed others in their places who would have been obedient." See H. Heaton, op. cit., p. 58, who takes a dimmer view of Gott's success in putting the machine to work.

58. For a more detailed description, see Alexander Tilloch, *The Mechanic's Oracle, and Artisan's Laboratory & Workshop; explaining, in an easy and familiar manner, the general and particular application of practical knowledge, in the different departments of science and art* (London: Caxton Press, 1825), pp. 145–147.

59. Cosmopolitan reference in *Journal de Rouen et du département de la Seine-Inférieure* (1798): 157, #38, 8 Floréal.

60. Archives nationales (AN), F 12 652, Pierre Laurens Daly, *Mémoire sur la état actuel des manufactures en cotton en France*, Sept. 1, 1790.

61. AN, F 12 2195, letter from F. Bardel to the Minister of the Interior, n.d. but probably 1797.

62. Jeff Horn and Margaret C. Jacob, "Jean-Antoine Chaptal and the Cultural Roots of French Industrialization," *Technology and Culture* 39(4)(1998): 671–698.

63. Mabel Tylecote, *The Education of Women at Manchester University 1883–1933* (Manchester: Manchester University Press, 1941), pp. 4–6. The founder of Owens College in Manchester stipulated in his will that only men were to be educated there. It took an act of Parliament to break the will. Note there were a few women present at the Leeds Philosophical and Literary Society in the 1820s.

64. AN F 12 2204, Dubois, "Le Conseiller d'état, Préfect du Département de la Gironde à ses Concitoyens," Summer 1801.

65. Robert Fox and Anna Guagnini, *Laboratories, Workshops, and Sites:*

Concepts and Practices of Research in Industrial Europe, 1800–1914 (Berkeley, Calif.: Office of History of Science and Technology, 1999), p. 14. For a prize established in the year 1800 in Lyon, see AN F 12 2359.

66. AN, F 12 2200, Fauchat, *État des machines à vapeur importées d' Angleterre en France depuis 1816*, dated April 7, 1819. For an overview of French industry in the period, see Gérard Béaur, Philippe Minard, and Alexandra Laclau, *Atlas de la Révolution française*. Vol. 10: *Économie*. Éditions de l'École des Hautes Études en Sciences Sociales (Paris, 1997).

67. *Bulletin de la Société d'Encouragement pour l'Industrie nationale,* a report by M. le baron de Prony dated September 13, 1809, and found in AN F12 2200.

68. Bibliothèque de la ville de Lyon, MS 5530, la Société libre d'Agriculture, histoire Naturelle & Arts utiles de Lyon; the range of the society was both agricultural and industrial, commencing in the year 1798.

69. AN, AD VIII 29, "Classification des places d'Elèves."

70. AN F 12, 533, Ministry of the Interior, "Rapport à Sa Majesté l'Empereur et Roi," November 23, 1808.

71. Patricia Jones, "Are educated workers really more productive?" *Journal of Development Economics* 64 (2001): 57–79.

72. Antoine Léon, "Promesses et ambiguités de l'oeuvre d'enseignement technique en France, de 1800 à 1815," *Revue d'histoire moderne et contemporaine* 17(3)(1970): 846–847.

73. AN, Roederer MSS 29 AP 75, f. 393: a *lycée* for 150 would have 9 professors and 3 administrators; f. 397: every district was to set up its own primary school; directive of April 5, 1802 (f. 399) said that mathematics was to be taught in secondary schools. Government would pay for students who were smart enough to secure a place; instruction was to include mathematics, pure and applied, and experimental physics, chemistry, natural history, statistics, and technology. There were to be two professors of science, one of physics and one of chemistry, as well as a professor for applied mechanics, *arts et metiers,* and technology in 1803. The goal was for 6,000 students in the *lycées*, 3,000 chosen by the government from the children of military and functionaries "who serve the republic well"; the other 3,000 were to be chosen by exam. A six-year course of study was to be instituted

and the government could distribute its largess unequally. Eventually La Fleche and one other of the old colleges was added and 6,400 pupils became the goal; f. 429: "le nombre d'eleves que doit avoir chaque lycee doit varier." It must be remembered that the state "ne seul qu'une prime pour former les colleges; et ce systeme actual peut eu quelque sorte se comparer au systeme du manufactures, Un Departement n'a't-il point de manufactures?" After further justifications for why the government should favor manufacturing, the report concludes that by age 15 or 16, the pupils would be nearly finished and would be studying mechanics and optics (see ff. 645) Professors were to use books approved by the government, which would consult Delambre and Cuvier at the Institute for advice.

74. Archives departementales du Nord (hereafter AD), IT 407 (printed brochure from 1820), Université de France, Collège Royal de Douai: "(3) Les objets de l'enseignement sont: la religion, les langues anciennes et modernes, les belles-lettres, la philosophie, les mathématiques, la physique, la chimie, l'histoire, la géographie, l'écriture, le dessin. Il y a un cours spéciale d'Anglais, dont le professeur est payé comme ces des cours précédens, par le Collège, et un cours d'Allemand, dont le Professeur reçoit le rétribution des élèves qui le suivre . . . Les élèves sont initiés à toutes les connaissances littéraires et scientifiques, indispensables pour être admis à l'école polytechnique, ou à toute autre école spéciale. Outre les treize Professeurs chargés d'enseignement, il y a un maître d'étude, ou répétiteur, par vingt-cinq élèves, chargé de les aider dans leurs études, de surveiller leur travail et de faciliter leurs progrès. Il y a un cabinet de physique, riche en instrumens, et un laboratoire de chimie bien organisé, pour que les élèves puissent, dans les sciences naturelles, joindre la pratique à la théorie. Ces ressources sont d'autant plus utiles, qu'une ordonnance royale prescrit que les candidats au baccalauréat seront examinés sur tous les objets de l'enseignement donné dans les Colléges Royaux et y comprix les mathématiques et la physique. Les élèves qui désirent prendre la grade de Bachelier, sont particulièrement exercés."

75. AD du Nord, MS IT 19/1, Facultés des sciences/Cours de physique à Lille, 1817–1852. "Ministre de l'Intérieur L'Etablissement d'un Cours de physique expérimentale à Lille est approuvé Paris, le 15 8 bre 1817." For salary, see MS1T 30/1.

76. AD du Nord, MS L 4841 from the year 1800.

77. AD du Nord, L 4842, and from the same period: "Il seroit difficile de ne pas sentir l'avantage d'un plan d'éducation aussi vaste et ainsi coordonné; il n'est presque pas un art, pas une profession utile et honorable, dont les connoissances spéciales ne dérivent de quelques-unes des sciences dont on vient de tracer le tableau: il sera aisé d'appercevoir que le cours de dessein, réuni aux cours de mathématiques et de physique, renferme tous les élémens de l'art de l'ingénieur, tant civil que militaire; d'artilleur, d'architecte (les jeunes gens qui se seront distingués dans ces sciences, ont la perspective d'être appellés à l'école polytechnique, d'où ils ne sortent que pour remplir des postes importans que le gouvernement leur confie); que le cours d'histoire naturelle, de physique et de chimie servent d'introduction aux états d'officiers de santé de toutes les classes, et que la chimie conduit à la perfection des procèds employés dans les manufactures, telles que les blanchisseries, les tanneries, dans l'art des teinturiers et des salpêtriers, etc. que les cours de grammaire générale, de belles-lettres, d'histoire, et de législation forment des hommes de loi, etc. Enfin il est clair que toutes les classes de la société doivent retirer un profit plus ou moins direct de l'ensemble des connoissances présentées à la jeunesse dans cet établissement, placé d'ailleurs sous l'influence de dix professeurs qui consacrent tout leurs temps aux différentes branches qu'ils enseignent."

78. AD du Nord, MS 2T 1208 Enseignement Secondaire et primaire, Généralités, 1812–1852, Rapports d'inspection en executant au decret du 15 novembre 1811: 1812–1813, Académie de Douai, L'Inspection à Monsieur le Recteur de l'Académie, Hazebrouck, 6 juin 1813, No. 1 Collège d'Armentières. "Les classes des Mathématiques composée de 7 élèves est extremement faible surout quand on considère qui M. Piette a été professeur dans une école centrale et dans deux lycées. Il paraît condomné à une longue médiocrité; on ne gagne guère à son âge; les meilleurs élèves de cette classe seront peut être bons à noter une autre année." Académie de Douai, L'Inspection à Monsieur le Recteur de l'Académie, Hazebrouck, 11 juin 1813, No. 3 Collège de Bailleul: "on reclame l'enseignement des mathématiques comme indispensables et comme devant faire fleurir le collège; c'est le voeu de toute la ville, on le demande pourquoi le Collège de Bailleul à trois Régents de latinité, lorsque celui d'Armentière qui est d'une tout

autre importance, n'a que deux régens de Latinité qui suffisent au Service plus un régent de Mathématiques."

79. Archives departementales, Seine-Maritime, MS XIX H 4, circulaires et instructions officielles relatives à l'instruction publique, 1802–1900.

80. AN F 4 1246 for the list and budgets.

81. AN F 14 11057, Ecole polytechnique: "sommaire des léçons du cours de mécanique analytique . . . par M. de Prony."

82. AN F 41246, Project de Réglement, article 24.

83. See AN F* 4 1847. For Chaptal's disillusionment with the Ecole Polytechnique, see Jean Pigierie, *La Vie et l'oeuvre de Chaptal* (Paris: University of Paris, 1931), p. 262.

84. Quoted in Marcela Efmertová, "Czech Technical Education: The Educational Reforms of Franz Gerstner and His Relationship with the Paris Ecole Polytechnique," *The Journal of the International Committee for the History of Technology* 3 (1997): 211.

85. AN F 4 2138 Ecole gratuite de dessin, Conservatoire des arts et métiers, 1816–1828. In 1823 4,182 francs was spent on machines as opposed to 343 francs for books; and in 1824 5269.15 francs on machines vs. 370 francs on books.

86. AD Seine-Maritîme, MS IT 1641, the library of the *lycée* in Rouen, 1810.

87. AN F 17 6770, Ecole centrale des arts et manufactures, création, 1828–1846.

88. See John R. Pannabecker, "School for Industry: L'Ecole d'Arts et Métiers of Châlons-sur-Marne under Napoleon and the Restoration," *Technology and Culture* 43 (2002): 254–256.

89. AN F 14 11057, f. 26, a remarkable document dated Paris, September 25, 1839.

90. M. Pinault, *Traité de Physique* (Paris: Gaume, 1839).

91. Hermione Hobhouse, *The Crystal Palace and the Great Exhibition: Art, Science and Productive Industry* (New York: The Athlone Press, 2002), p. 194.

92. Auerbach, *The Great Exhibition,* p. 32.

93. Auerbach, *The Great Exhibition,* pp. 28–29.

94. *The Art Journal: Illustrated Catalogue; The Industry of All Nations,* 3 vols. (London: George Virtue, 1851), p. I.

95. *Official Descriptive and Illustrated Catalogue by Authority of the*

Royal Commission (London: Spicer Bro., 1851), p. 3 in the words of Prince Albert.

96. Ibid., p. 4.

97. See, for example, *The Art Journal: Illustrated Catalogue.*

98. Ibid., p. 85.

99. Ibid., pp. 86–87.

100. Charles Tomlinson, ed., *Cyclopaedia of useful arts, mechanical and chemical, manufactures, mining, and engineering* (London: G. Virtue, 1852–1856).

101. Daniel Drake, M. D., *An Anniversary Discourse on the State and Prospects of the Western Museum Society: Delivered by appointment, in the Chapel of the Cincinnati College* (Cincinnati, Ohio: For the Society, 1820), pp. 32–33. See also Timothy Claxton, *Memoir of a Mechanic* (Boston, 1839), pp. 61, 82, 104–32.

Epilogue

1. Leeds University, Brotherton Library, Special Collections, MS Dep. 1975/1/5 (Box 2) Council Minute Book, 1819–1822, Papers of the Leeds Philosophical and Literary Society.

2. Archives d'Etat, Liège, Fonds Français Prefecture, inv. nr. 452–454. Letter of April 2, 1808, from the professor of mathematics and physics in Liège, Vanderheyden, to the Bureau of Administration in Paris requesting money for a laboratory, chemical samples, and a small mineralogical collection: "Vous connoissez trop bien, Messieurs, l'utilité d'un cours de chimie et de minéralogie pour le progrès des arts chimiques et manufactures en ce département."

3. The printed *Programme des cours de L'École Centrale du département de l'Escaut, qui s'ouvriront le primier brumaire an XII,* Ghent, 1802, pp. 6–7, and found in Archives nationales, Paris, F17 1344 14.

4. Rijsarchief Gent, Hollands Fonds, inv. nr 611/2 for details on the exposition.

5. Rijksarchief Liège, 03. 01 inv.nr 2523, 2624.

6. Quoted in Maria M. Portuondo, "Plantation Factories: Science and Technology in Late Eighteenth-Century Cuba," *Technology and Culture* 44 (April 2003): 246.

7. See Robert Fox, "Science, Practice and Innovation in the Age of Natu-

ral Dyes, 1750–1860," in Maxine Berg and Kristine Bruland, eds., *Technological Revolutions in Europe* (Northampton, Mass.: Edward Elgar, 1998), pp. 86–95. The essay draws too sharp a distinction between theory and practice.

8. See Joel Mokyr, *The Gifts of Athena: Historical Origins of the Knowledge Economy* (Princeton, N.J.: Princeton University Press, 2002); for the argument about the takeoff being after 1800, see R. Bin Wong, *China Transformed: Historical Change and the Limits of European Experience* (Ithaca, N.Y.: Cornell University Press, 1997).

Acknowledgments

The authors wish to thank the staffs of various libraries: the Young Research Library, UCLA and the Clark Library, the Bakken Library in Minneapolis, The British Library, the Scottish Record Office, The Brotherton Library in Leeds, and various French, Belgian, and Dutch departmental archives in Rouen, Lille, Liège, Ghent, and Middelburg in Zeeland. The *Canadian Journal of History* allowed a reprinting of a few pages of an article on Manchester that now forms a part of Chapter 5. The coauthor, David Reid, kindly consented to the reprinting. So too a few paragraphs from "A Women's Scientific Society in the West: The Late Eighteenth-Century Assimilation of Science" appeared in *Isis* (2003), and we wish to thank that journal and the coauthor, Dorothée Sturkenboom, for permission to reprint. Research assistance was provided by Mindy Rice, Tami Sarfatti, and Eric Casteel. Jeff Horn and Lynn Hunt kindly read portions of this book, as did the members of the European History Colloquium at UCLA. The authors wish to thank the other editor of the series, Spencer Weart, and the Harvard University Press editors. MCJ gratefully acknowledges the support of the National Science Foundation grant no. 9906044.

Index